INTERNET OF THINGS

An Easy Hands-on Guide

INTERNET OF THINGS
An Easy Hands-on Guide

V. Jeyalakshmi
Professor
Electronics and Communication Engineering
Anna University
Chennai
Tamilnadu, India

TECHSAR™
NEW DELHI

Published and Printed by

Techsar Pvt. Ltd.
4435-36/7, Ansari Road, Daryaganj
New Delhi-110 002 (India)
E-mail: tech_info@techsarworld.com
Website: www.techsarworld.com

ISBN: 978-93-90620-48-7

PREFACE

The aim of this book is to empower the reader to get sufficient knowledge about Internet of Things (IoT) starting from basics. As an easy hand-on approach guide, the chapters are organized in such a way that each chapter explores with layer-wise explanation with corresponding exemplary. A thirst to prepare this hands-on guide arose to share the knowledge of emerging technologies among professional, and non-professional technical students. The content coverage will fulfill the need of all technical researchers, faculties and students' expectations. In addition to the new technological representations the existing technology is explained with proper definitions and its drawbacks. Then, how the problem overcomes by the proposed system is given. So this book is fruitful for the young researchers in developing as well as implementation stage. Various case studies with the proper codes is the further add-on feature of this book.

V. Jeyalakshmi

CONTENTS

PART-I
Introduction & Concepts

1

COMMUNICATION ERA

1.1 INTRODUCTION

The word **"COMMUNICATION"** comes from the Latin language which literally means "makes something common" by action of communicating. The human communication evolved from the stone-age era. All the living things in the universe have some system of the communication especially human beings have some special talent of the communication through the spoken language. The four transitions of human communication from the ancient periods are:

- **Petroglyph:** It is a method of carving on a rock surfaces. It shows the events that happened in the ancient times.
- **Pictograms:** The ideas are exploited by means of pictures. It narrates the real stories about the event in chronological order.
- **Ideograms:** The concept of the events was narrated in ideograms. From this people got some ideas related to the event.
- **Writing:** Initially the ancient writing was logographic in nature. It was fully based on the above methodologies. Then from the ideograms, the humans started to communicate using the alphabet writing representations with symbols.

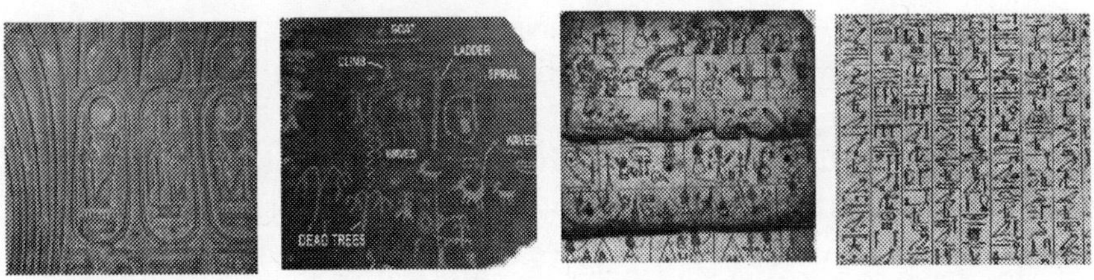

(a) Petroglyph (b) Pictograms (c) Ideograms (d) Logograms

Fig. 1.1 Pictures representing communication eras

The different communication eras and their symbol representations are shown in Figure 1.1. Naturally, humans having some form of speaking skill like voice signal, once they started to write then they thought innovatively and started to communicate verbally, this is a milestone of the human communication. From 1500 BC to 1800 BC information sharing was eventually improved with the invention of printing. Then the mailing (posting) system started. People shared their information by posting letters from one place to another. Later on telecommunication took place. Telecommunication evolution is represented in Figure 1.2.

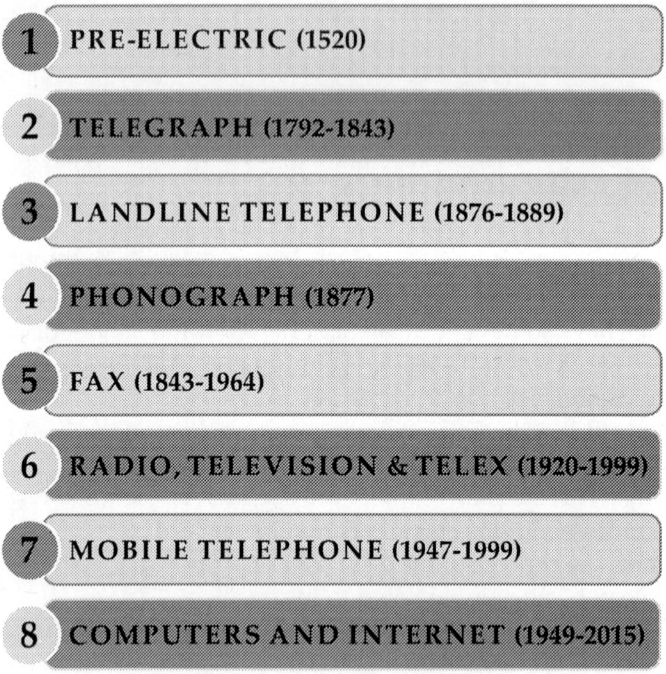

Fig. 1.2 Evolution of telecommunication

Telecommunication begins with the term pre-electric signals in 1520, i.e., the passing of signals by the reflection of sunlight from the metal plate. Then they communicated with the telegraphic method which was established for long distance transmission in the year of 1792 to 1843. Further, the generation evolves with the invention of landline telephonic communication in 1876. From 1920 to 1999 radio, television and TELEX power line carrier communication via electric towers were in use. In the earlier period of 20th century radio and television communication started, until the mid of this century the whole world communicated using wired transmission technique successfully. At the end of 20th century the idea of digital communication arose and the concept of Internet started to emerge.

In the 21st century, information technology started to boom significantly. Let us view what are the various aspects that have already happened in technology and things to happen in future.

1.2 CURRENT COMMUNICATION TECHNOLOGIES

1.2.1 Information Age

Information age is one of the important periods in 20th century with rapid development of technology. Based on the evolution, telecommunication was started with analog communication. The voice-like analog signals were transmitted after a different variety of analog modulation (AM, FM & PM) for long distance communication to strengthen the signal. The signals were transmitted initially in simplex mode (one way) then developed as half duplex (any one) and then standardized with full duplex. The communication path was a single line, supported to carry both receiver and transmitter signals at the same time. So the analog communication of voice and data through wired and wireless reaches the people successfully to develop their innovations in education, industry and business. Now the technology turns to think about the digital communication and how to make it more reliable and fast. The analog signals are converted into digital signals then modulated with digital modulation like shift keying techniques, pulse code modulation and delta modulation. Instead of manpower to do work most of the works are machinated (automated) in this age. The demand of digital need has increased in this computerized period. So, the digital signal conversion was made to be compactable to all electronic gadgets. Digitalization of communication provides more user-friendly approach with more storage capacity. Because of automation numerous data is needed to be captured for analysis and taking decision. The development of various storage devices and new methodologies were discovered to provide the automation and high speed computation. With this improvement in technology, storage of data becomes larger and larger. The requirement of more memory capacity, in turn, increases the cost. To overcome these issues the Cloud is used to back up more information by pushing data into the Cloud. So now, the digital technology changes the slow speed communication to high speed communication and also for both automation and digitization by the Internet of Things. Nowadays, it started to merge every "things" in the real world to Internet and trying to automate with digitization. This is the brief history of communication from earlier to current stage. After a brief refresh from ancient to earlier communication technology let's see into the current smart technology Internet of Things.

1.2.2 Internet of Things (IoT)

Definition

IoT is defined as connecting all the real-world electronically embedded devices through the Internet and it can be accessed by any authorized person at any time and on any place. The action of this ubiquitous computing turns every ordinary operation into the smart work.

Example: Smart Home, Smart Cars, etc.

1. All the electronic devices are interconnected in smart home. Recent home electronic devices are automated with smart phones also. Consider an example Smart Air Conditioner. Air conditioning units are normally operated by remote or manual mode or by using smart phones. But for Smart AC, all the operation controls are interconnected through Internet, and can operate on or off and can regulate it from anywhere with the support of the IoT devices or software.
2. Electronic bulbs connected to Internet regulates the light illumination power on or off mechanism by means of smart IoT unit.

General IoT Structure

IoT devices consist of various embedded software in hardware devices. Hardware devices contain various accessories and IoT devices. Software contains various application based platforms and services. Figure 1.3 shows a brief idea of general IoT structure.

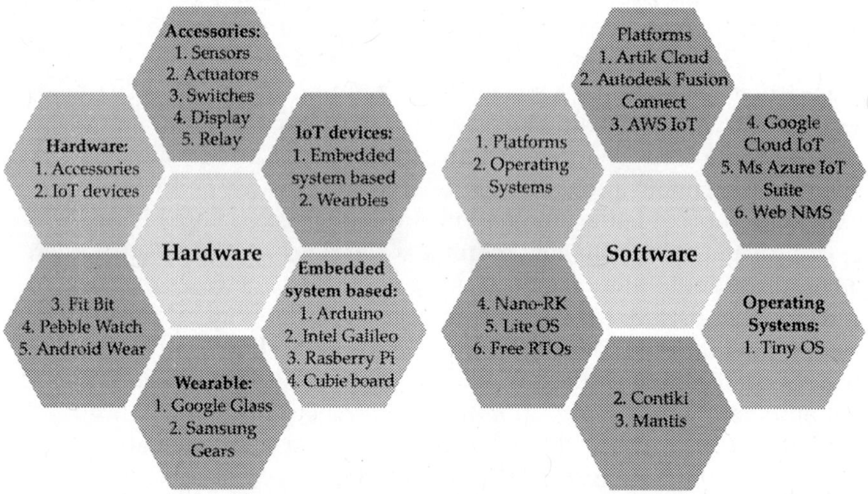

Fig. 1.3 General structure of IoT

1.2.2.1 Characteristics of IoT

The characteristics of IoT are very important to the designer or project developer to fulfill the application constraint. Some of the characteristics are listed below.

1. **Artificial Intelligence:** In contrast to the conventional functions, the automatic functions are the latest trend nowadays. Usually, smart

devices come with the in-built software and hardware to perform certain operations. In order to make it smarter, the way of coding is advanced in such a way that machine will automatically find the needs of the user based of the history of their research.

Example: Spam filters, smart email categorization, text editors or auto correct, face detection and recognition.

2. **Scalability:** The system can be modeled with in-built memory and have the provision of storing data in Clouds for processing and analyzing large number of data. Various new technologies are available to process the data stored in the Cloud.

   Example: Google storage accounts, back up of documents and the photos in Cloud. (Depending upon the storage capacity which can be stored the data from smaller to the larger size up to terabytes.)

3. **Inter Communication:** It is possible to enable the Inter, machine to machine communication between various devices. In IoT the real-world devices or systems are connected to the Internet in order to enable public access and lots of private communication would also be enabled between the intermediate devices. Various types of intermediate communication are available as listed below.
   - human to machine communication,
   - machine to human communication,
   - machine to vehicle communication,
   - vehicle to vehicle communication, and
   - machine to machine communication.

   Example:
   1. Smart mobile phones are used to switch on TV, fan and AC and also to control remote plants like reservoir, etc.
   2. Voice message to Amazon echo which connects to Alexa to play music, make calls, check news, etc.

4. **Change of the State Dynamically:** The device can provide response and access at any time that dynamically changes the service. Consider any one of the smart communication systems to evaluate its performance. If the response time for particular action is too long, then the machine is not a smart machine. The important attitude for the entire smart machine is to provide the dynamicity of state at any time.

   Example: Consider any measurement whose sensors will be turning on and off dynamically when it is necessary to collect the data.

5. **Security:** Security and privacy is an important characteristic that is necessary to provide integrity and authentication. Here the device can be modeled with the high level security and authentication.

Without proper privacy and the security, anyone can access and get everyone's information in IoT.

Example: Insecure data transfer and storage in medical and industrial field, security problems in device update management, industrial espionage, eavesdropping, etc.

1.2.2.2 *Features of IoT*

Some important features of IoT are listed below:

(a) **Connectivity:** Connectivity refers to ascertaining correct affiliation between all the items connected in IoT platform. When connecting the IoT devices, it desires a high speed electronic messaging between the devices and the Cloud to alter reliable, secure and bidirectional communication.

(b) **Analyzing:** After connecting all the relevant devices and details, it involves the period of time analyzing information collected to make effective business intelligence.

(c) **Integrating:** The IoT consists of more applications and devices integrated in a single device. Its integration with the varied models is to boost the user expertise further.

(d) **Artificial Intelligence:** The IoT makes things sensible and enhances life through employing information. For instance, assume a coffee machine whose beans are about to finish, then the coffee machine itself order for beans from the retail merchant.

(e) **Sensing:** The IoT technology brings passive networks to active networks. The detector devices are employed in IoT technologies to observe and update any modification within the surroundings and report to the controller.

(f) **Active Engagement:** The IoT makes the active engagement between the various devices in connected technology, products and services.

(g) **Endpoint Management:** It is necessary to be the termination management of all the IoT system otherwise. For instance, if a coffee machine itself orders the low beans level once it goes to finish however what happens once it orders the beans from a retail merchant and that are not receiving within a couple of days. It ends up in the failure of the IoT system. So, there should be a requirement for termination management.

1.2.2.3 *Challenges of IoT*

Expansion of technology smartness in IoT, leads to many challenges also. The important challenges are listed.

(a) Power management
(b) Data connectivity
(c) Hardware compatibility
(d) Security
(e) Delivering value to the customer

(a) **Power Management:** Power management is an often overlooked yet critical area for both consumer and industrial IoT devices. In a battery powered device, optimizing dynamic as well as static power is imperative. Power optimization can be addressed in three ways:
- Power management control
- IP implemented for low power
- Power aware software

(b) **Data Connectivity:** One of the crucial requirements for the success in the IoT technology is the availability of uninterrupted, reliable as well as bidirectional signaling which should be enough for collecting as well as routing all data between two or more devices. This can be achieved by means of certain cellular communication optimization.

(c) **Hardware Compatibility:** Data capturing mainly happens through varied sensors, PLCs, etc., that are area unit connected to IoT gateways to gather and transmit knowledge to the Cloud. Enterprises ought to meticulously establish the instrumentality, hardware, and existing heritage machines supporting their goals and business outcomes. Once the area unit heritage machines unavailability of same PLCs and sensors concerned, then the IoT implementation challenge becomes additional crucial. Adding external sensors to the heritage machines is one fast work around, however it will not be a foolproof, creating a really difficult task. Therefore, distinguishing the physical devices and understanding the associated compatibility problems before IoT implementation is suggested.

(d) **Security:** With occurrences of the many malware attacks recently, clients in the industrial area unit are apprehensive regarding knowledge security. Hence, IoT service suppliers ought to make sure that their knowledge goes to be safe. These knowledge security problems will be taken care by employing a comprehensive governance mode that provides secure access to sensitive reports and knowledge. This makes always to ensure the security awareness in all IoT applications.

(e) **Delivering Value to the Customer:** Defining clear statement helps for the success of Associate in Nursing IoT implementation. However, this is often what most IoT service suppliers missed out. They must perceive the solutions of delivering value will impact the potency, client satisfaction and productivity within the end of the day. This complete cycle desires nice retrospection and there can't be an even bigger IoT implementation challenge than this gap of understanding the client downside statement. Therefore, it's crucial for IoT consultants to figure out the key performance indicators to give an improved IoT resolution.

1.3 IoT ENABLING TECHNOLOGIES

IoT supports both wired as well as wireless communications. Wired communication for longer distance is a cost effective. Wireless communication for larger data is mind effective. In order to enable this, five major technologies are available. They are:

1. Fog computing (acts like a gateway near to the user)
2. Cloud computing (can process and store data in GB)
3. Big data analytics (can process and store data more than GB)
4. Machine learning and
5. Block chain

Figure 1.4 represents the technology.

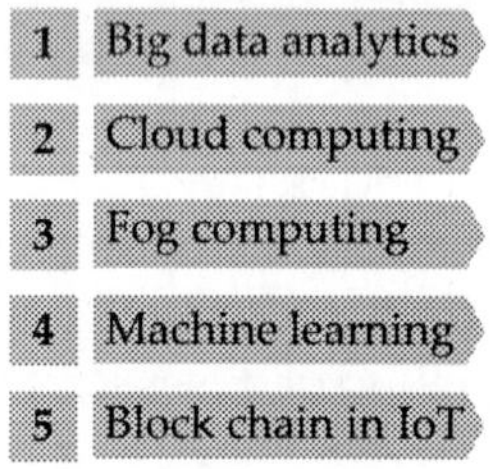

Fig. 1.4 IoT enabled technologies

1.3.1 Big Data Analytics in IoT

What makes massive amount of data with necessary quality to businesses is that it makes it able to extract analytics and consequently information, by that a business will succeed with competitive advantage. There are some platforms for massive knowledge analytics like Apache Hadoop and SciDB. However, these tools are hardly sturdy enough for massive knowledge needs of IoT. The quantity of IoT knowledge usually is too large to be fed and processed by the accessible tools. In support of the IoT, these platforms ought to work in a period of time to serve the users expeditiously. For example, Facebook has used associate in nursing improved version of Hadoop to research billions of messages per day and provide period of time statistics of user actions.

1.3.2 Cloud Computing in IoT

Cloud computing (CC) offers a new management mechanism for enormous knowledge that allows the process of (data or information) and also the extraction of valuable information from it. Using CC for the IoT is not a straightforward task.

- **Synchronization:** Synchronization between totally different Cloud vendors presents a challenge to supply period of time services since services area units are designed on prime of assorted Cloud platforms.
- **Standardization:** Standardizing Cloud computing additionally presents a big challenge for IoT Cloud-based services due to interoperation with the varied vendors.

- **Balancing:** A balance between general Cloud service environments and IoT needs, presents another challenge.
- **Reliability:** Security of IoT Cloud based services presents another challenge due to the variations in the protection mechanisms between the IoT devices and the Cloud platforms.
- **Management:** Managing Cloud and IoT systems is additionally a difficult issue thanks to the very fact that each one has totally different resources and elements.
- **Enhancement:** Validating IoT Cloud based services is important to confirm providing smart services that meet the purchaser's expectations.

1.3.3 Fog Computing in IoT

Fog Computing (Cloudlets or edge computing) acts as a bridge between sensible devices and large-scale Cloud computing and storage services. Through fog computing, it's possible to extend Cloud computing services to the edge devices of the network. Owing to their proximity to the end-users compared to the Cloud data centers, fog computing has the potential to provide services that deliver higher delay performance. It ought to be emphasized here that, generally there's a big difference in scale between the fog and also the Cloud such that the Cloud has large process, storage and communications capabilities compared to the fog.

1.3.4 Machine Learning in IoT

The focus of machine learning is to insist the ability of automatic work in all system without depending on any specifically designed program. As a subset of artificial intelligence, machine learning learns and improves itself via system experience without the human intervention, the ability of providing the artificial decision making by the machines seems to be the stepping stone of success for IoT. In general, machine learning was evolved from pattern recognition and computational learning theory. The learning theory usually uses supervised, unsupervised and reinforcement methods for data detection and analysis. Among the three methods, supervised and unsupervised are used widely for IoT smart analysis. For data processing tasks in smart analysis classification, regression, clustering, feature extraction and anomaly detection methods are used. Some of the machine learning algorithms are for IoT use cases given in Table 1.1.

1.3.5 Block Chain in IoT

Block chain was first explained by Stuart Haber and W. Scott Stornetta in their research paper titled "How to time stamp a digital document" published in the *Journal of Cryptography* in 1991. Block chain is a document full of entries that are shared with a group of people and organization. It

Table 1.1 Algorithm for machine learning

Machine learning algorithm	IoT smart city use cases	Metric to optimize
Classification	Smart traffic	Traffic prediction, increase data abbreviation
Clustering	Smart traffic, smart health	Traffic prediction, increase data abbreviation
Anomaly detection	Smart traffic, smart environment	Traffic prediction, increase data abbreviation, finding anomalies in power dataset
Support vector regression	Smart weather prediction	Forecasting
Linear regression	Economics, market analysis, energy usage	Real-time prediction, reducing amount of data
Classification and regression trees	Smart citizens	Real-time prediction, passengers travel pattern
Support vector machine	All use cases	Classify data, real time prediction
K-nearest neighbors	Smart citizen	Passengers' travel pattern, efficiency of the learned metric
Naive bayes	Smart agriculture, smart citizen	Food safety, passengers travel pattern, estimate the numbers of nodes
K-means analyze	Smart city, smart home, smart citizen, controlling air and traffic	Outlier detection, fraud detection, small data set, forecasting energy consumption, passengers travel pattern, stream data analyze
Density-based clustering	Smart citizen	Labeling data, fraud detection, passengers travel pattern
Feed forward neural network	Smart health	Reducing energy consumption, forecast the states of elements, overcome the redundant data and information

is a decentralized network. A block is a unit containing a set of confirmed transactions. Block chain is a game changer of IoT. It provides a perfect solution for the safer and secure transmission and reception of data in IoT environment. Block chain acts as a private ledger to collect all the information from the sender and transmit it to the receiver without having access from the central server. If the block chain is considered a ledger, then think the block is as a page or book. It enables all its IoT business partners to get tamper-resistant records. Building trust between every IoT transactions, reducing transaction cost by removing intermediaries, and allowing instantaneous transactions are the three important benefits of block chain. To illustrate the above benefits of block chain, IBM listed out logistics and trade lanes examples. In that the benefits are reflected in terms of frequent auditing, perfect accountability, and providing new forms of contracts and speedier transactions. Public, private and hybrid block chain of various types of block chain are available. Having many advantages, block chain in IoT also faces some challenges like scalability, interoperability, legal compliance and regulations. Currently, all industries are working towards it to resolve all the problems for providing a perfect IoT environment. Some of the significant block chain impact industries are supply chain and logistics, automotive industry, smart homes, sharing economy, pharmacy industry, agriculture and water management. It provides efficient 24 × 7 transaction process in secure way and improves efficiency of the system.

1.4 SUMMARY

In this chapter a detailed explanation on communication era development with its recent technologies like IoT, its role, characteristics, features and challenges on communication were discussed. A brief description on the IoT enabling technologies towards big data analytics, Cloud computing, Fog computing, machine learning and block chain for IoT applications were given.

REVIEW QUESTIONS

1. Describe the technological gap in telecommunication.
2. Define IoT with suitable example.
3. Describe the general structure of IoT.
4. What are the characteristics of IoT?
5. What is the role of artificial intelligence in IoT?
6. Define endpoint management in IoT.

7. Write the challenges of IoT.
8. List some IoT enabling technologies.
9. Compare Fog computing and Cloud computing in IoT.
10. Describe Big data analytics and role of block chain in IoT.
11. Write some applications of IoT.
12. What is the need of IoT? When to prefer IoT based applications?

2
IoT DEVICES

2.1 INTRODUCTION

This chapter introduces various IoT devices used for IoT applications, IoT devices with pins and signals. It also elaborates the characteristics of IoT sensors and their working principles. Along with that how IoT devices play a vital role to develop smart devices or system for different field of applications and also discusses how it varies from conventional system.

2.2. WIRELESS SENSORS

Sensors are used to detect and respond for the physical changes in the form of electrical and optical signals. The physical parameters like temperature, blood pressure, speed are converted into the electric signal and measured using sensors. Before discussing about some examples of sensors it is necessary to consider certain features. They should have accuracy, resolution, environmental condition, cost, range, repeatability and calibration. Various kinds of sensors are used in industries. Internet of things has taken the evolution of sensors in different level. In IoT network, the sensors collect data, pushes and shares it with the whole network of connected devices. The different sensors are listed which are more frequently used for IoT applications. They are explained as follows.

(a) Temperature Sensor

It is a device that provides temperature measurement through the electric signal. In other words this sensor detects and measures the hotness and coolness and converts it into an electrical signal. The different types of temperature sensors are LM35IC, thermistor, thermocouples, RTD (resistor temperature detector). Figure 2.1 gives the model of temperature sensor along with pin description in Table 2.1.

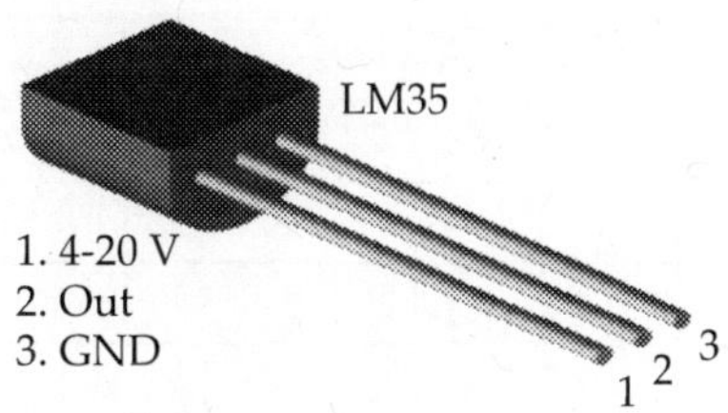

Fig. 2.1 Temperature sensor

(https://www.researchgate.net/Fig./Temperature-Sensor-LM35_fig3_323905602)

Table 2.1. Pin description of temperature sensor

Pin	Pin name	Description
1.	VCC	Input voltage is +4 V to 20 V for typical applications
2.	Analog out	There will be increase in 10 mV for raise of every 1°C. Can range from –1 V (–55°C) to 6 V (150°C)
3.	Ground	Connected to ground of circuit

Uses

Temperature sensors are used to measure temperatures in computers, mobile phones, automobiles, air conditioners, industries, etc.

(b) Infrared (IR) Sensor

IR sensor is an electronic device that emits IR waves to perform sensing operations based on some features of the surroundings. It also measures the heat of an object as well as detects the motion of an object. Figure 2.2 gives the model of IR sensor along with pin description in Table 2.2.

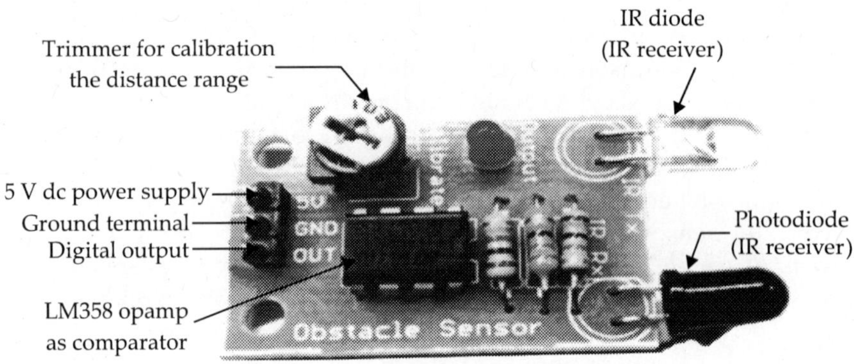

Fig. 2.2 IR sensor

(https://components101.com/sensors/ir-sensor-module)

Table 2.2. Pin description of temperature sensor

Pin	Pin name	Description
1	VCC	Power supply input +5 V DC
2	GND	Power supply ground
3	OUT	Active high digital output

Uses

Mobile phones, robots, industrial assembly, automobiles, remote control, etc.

(c) Proximity Sensor

This sensor detects the presence of the nearby objects without any physical contact. It operates with electromagnetic radiation and detects the changes in the reflected signal. Table 2.3 gives various types of proximity sensors. They are inductive sensor, capacitive sensor, photoelectric sensor and ultrasonic sensor. Figure 2.3 gives the model of proximity sensor along with pin description.

Table 2.3. Types of proximity sensor

Technology	Sensing range	Applications	Target materials
Inductive 	< 4-40 mm	Any close-range detection of ferrous material	Iron, steel, aluminum, copper, etc. 
Capacitative 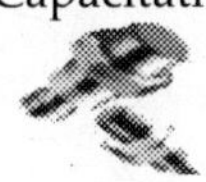	< 3-60 mm	Close-range detection of non-ferrous material	Liquids, wood, granulates, plastic, glass, etc. 
Photoelectric 	< 1 mm-60 mm	Long-range, small or large target detection	Silicon, plastic, paper, metal, etc. 
Ultrasonic 	< 30 mm-3 mm	Long-range detection of targets with difficult surface properties. Color/reflectivity insensitive	Cellophone, foam, glass, liquid, powder, etc. 

(https://www.machinedesign.com/automation-iot/sensors/article/21831577/proximity-sensors-comparison)

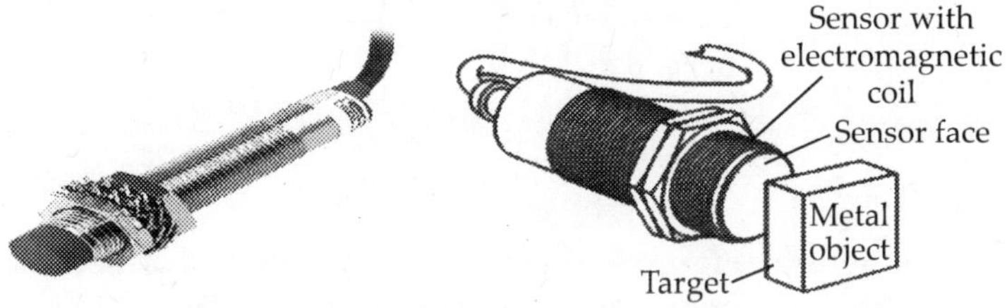

Fig. 2.3 Proximity sensor

Inductive Sensor

Inductive sensor (Figure 2.4) is a fast responding processes to detect an object. This is a non-contact sensor device which senses when the object is passing near to it. In this type of sensor, the range depends on the capacity and size of the device.

This method is mostly used in industrial applications to increase the production, quality by monitoring, etc. This is mostly used for metal detection applications. It helps in identifying the leakages, heavyweights, speed, etc. Inductive sensors are available in cylindrical and block type.

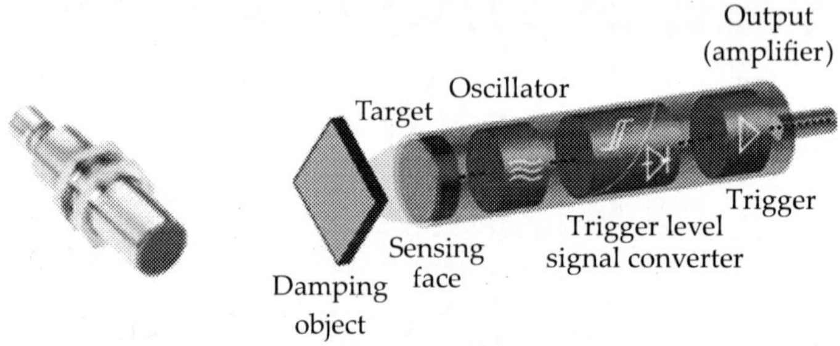

Fig. 2.4 Inductive sensor

(https://www.motioncontroltips.com/what-are-inductive-proximity-sensors/)

Capacitive Sensor

A capacitive sensor is used to identify both metal and non-metal objects. Capacitive sensor (Figure 2.5) works on the basis of capacitance. It works with the help of current flow which, in turn, gives change in capacitance. When the object is near the sensor it gets indication with the help of high current flow and is able to detect the presence of the object.

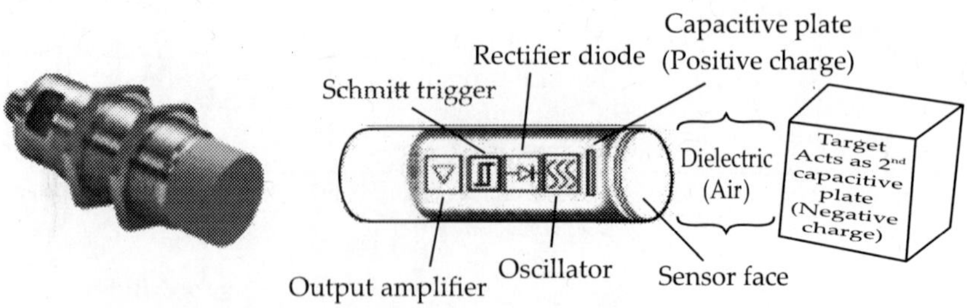

Fig. 2.5 Capacitive sensor

(https://expert.alliedelec.com/sensing/capacitive-proximity-sensors/)

Photoelectric Sensor

A photoelectric sensor (Figure 2.6) is used to identify the presence of an object with the help of a light source emitted through it. This light helps to identify the object in three ways: direct reflection of light, reflection with the help of reflector and with the help of through beam.

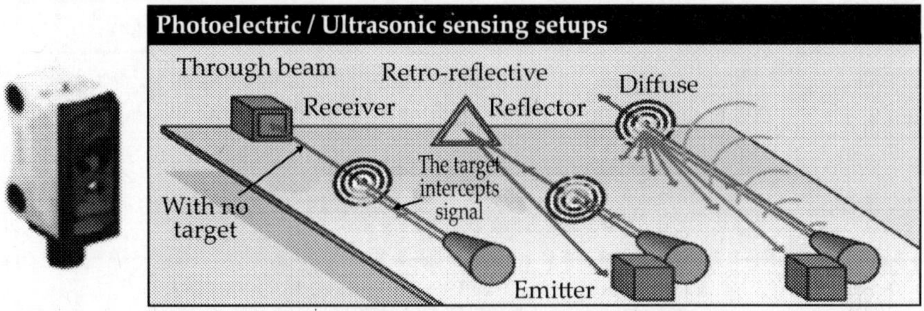

Lasers or sound waves serve as the signal in three setups. In through beam and retro-reflective, the signal shoots from the emitter ro receiver until the target cuts it off. In diffuse sensing, the signal diverges until a target moves in and reflects some back to the receiver.

Fig. 2.6 Photoelectric sensor

(https://www.machinedesign.com/automation-iiotphotoelectric-and-ultrasonic)

Ultrasonic Sensor

Ultrasonic device (Figure 2.7) is used to identify the presence of an object near it with the help of ultrasound waves. This type of sensor is used to measure high accuracy measurements. This type of sensor may affect environment. Due to this, it has an additional protection. Ultrasonic sensor is used in parking, automobile sensors, automatic systems, flow meters, etc. It cannot be used above 90°C of temperature and below –45°C. It measures the distance of an object using ultrasound waves. Ultrasonic sensors act as a transducer to send and receive ultrasonic pulses and relay back information about the objects proximity. The detailed pin description of ultrasonic sensor given in Table 2.4.

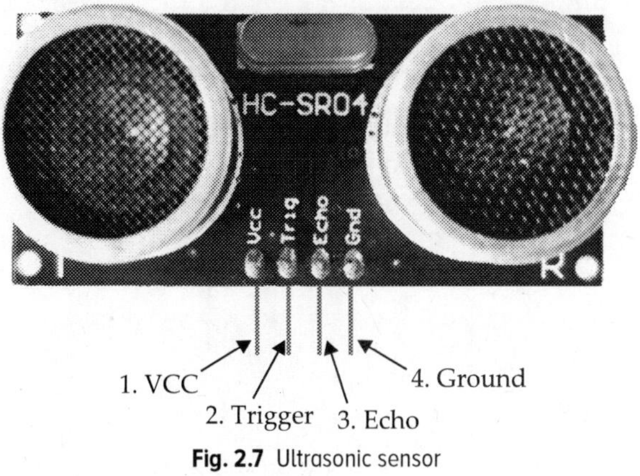

Fig. 2.7 Ultrasonic sensor

(https://components101.com/ultrasonic-sensor-working-pinout-datasheet)

Table 2.4. Pin description of ultrasonic sensor (Figure 2.7)

Pin	Pin name	Description
1	VCC	The VCC pin powers the sensor, typically with +5 V DC.
2	Trigger	Trigger pin is an input pin. This pin has to be kept high for 1 μs to initialize measurement by sending ultrasonic waves.
3	Echo	Echo pin is an output pin. This pin goes high for a period of time which will be equal to the time taken for the ultrasonic wave to return back to the sensor.
4	Ground	This pin is connected to the ground of the system.

Uses

It is mainly used for distance measurement and also to measure wind speed and direction, tank or channel fluid level through air or water. A proximity sensor is used in mobiles, vehicles, automatic lighting system, automatic water tap system, etc. It is also used to monitor machines'. Vibration by measuring the variation between shaft and support bearing, large steam turbines, and compressor. Motors are used as sleeve type bearings that use the proximity sensors.

(d) Pressure Sensor

This sensor is used for pressure measurement of gases or liquids. Pressure sensors measure valve in terms of force per unit area. It acts as a transducer by generating electric signal as a function of pressure imposed. Figure 2.8 gives the model of pressure sensor along with pin description in Table 2.5.

Fig. 2.8 Pressure sensor

(https://components101.com/sensors/bmp180-atmospheric-pressure-sensor)

Table 2.5. Pin description of pressure sensor

Pin	Pin Name	Description
1	VCC	Connected to +5 V DC
2	GND	Connected to ground.
3	SDA	Serial Data pin (I^2 C interface)
4	SCL	Serial Clock pin (I^2 C interface)
5	3.3 V	If +5 V is not present, can connect to +3.3 V pin

Uses

Pressure sensors are indirectly used to measure other variables such as fluid, gas flow, speed, water level and altitude.

(e) Level Sensor

Level sensors are used to determine the liquid fluid level that flows in a open system or closed system. There are two types of level measurement.

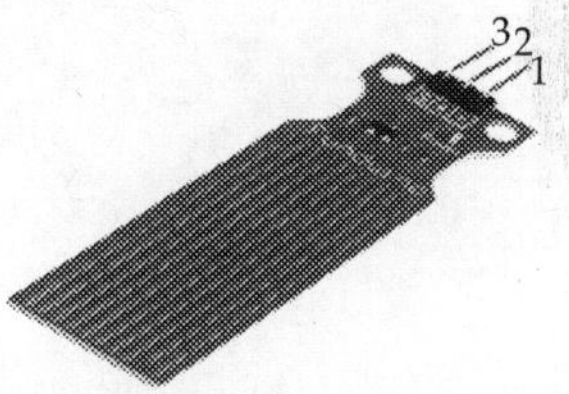

Fig. 2.9 Level sensor

(https://www.instructables.com/Methods-of-Detecting-Water-Level-Arduino/)

1. Continuous measurement. It measures level to precise limit and gives accurate result.
2. Point level measurement. It just measures the level of liquid, i.e., high or low.

Water level sensor is provided with three output terminals. Figure 2.9 gives the model of Level sensor along with pin description in Table 2.6.

Table 2.6. Pin description of level sensor

Pin	Symbols	Description
1	+	Power supply input +5 V DC
2	–	Power supply ground
3	S	Signal

(f) Smoke Sensor

This sensor senses the smoke typically as an indicator of fire. Commercial security device issues signal to a fire alarm control panel as a part of alarm system. It is also used as a household smoke detector. Figure 2.10 and its embedded module in Figure 2.11 along with the individual pin description in Tables 2.7 and 2.8.

Fig. 2.10 Smoke sensor

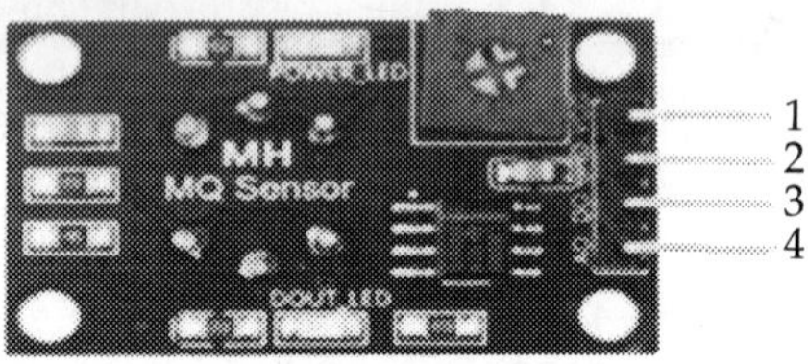

Pin No.	Pin Name
1	VCC (+5 V)
2	Ground
3	Digital out
4	Analog out

Fig. 2.11 Smoke sensor module

(https://components101.com/mq2-gas-sensor)

Table 2.7. Pin description of Smoke sensor circuit board

Pin	Pin name	Description
1	VCC	This pin powers the module, typically the operating voltage is +5 V DC (power supply +5 V DC).
2	Ground	Used to connect the module to system ground.
3	Digital out	To get digital output from this pin, by setting a threshold value using the potentiometer.
4	Analog out	This pin outputs 0-5 V analog voltage based on the intensity of the gas.

Table 2.8. Pin description of smoke sensor

Pin	Pin name	Description
1	H-Pins	Out of the two H-pins, one pin is connected to supply and the other to ground.
2	A-Pins	The A pins and B pins are interchangeable. These pins are tied to the supply voltage.
3	B-Pins	The A pins and B pins are interchangeable. One pin acts as output while the other is pulled to ground.

Uses

Mostly used for gas leakage detection in home, fire alarm, etc.

(g) Touch Sensor

Touch sensor (Figure 2.12) is sensitive to touch, force and pressure. It is also called tactile sensor. The working of touch sensor is similar to simple switch circuit. If the circuit is opened there is no current flow, if closed, then there is a current flow.

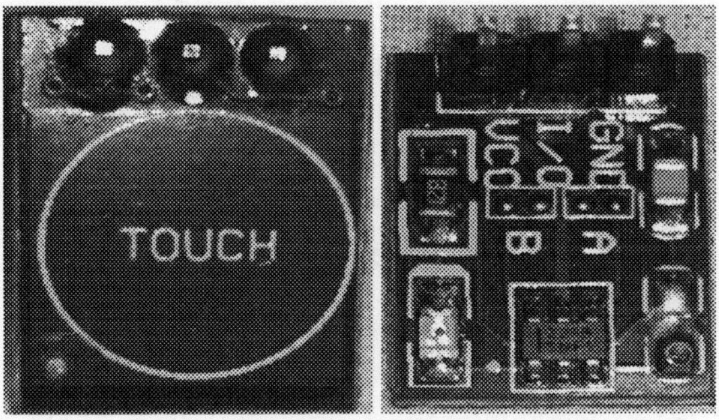

Fig. 2.12 Touch sensor

(https://www.electronicscomp.com/ttp223-1-channel-capacitive-touch-sensor-module)

Uses

Display applications, smart homes and industrial applications.

2.3 RFID

Radio frequency identification (RFID) (Figure 2.13) is a form of wireless communication which is used to identify an object using electromagnetic coupling in the radio frequency portion of electromagnetic spectrum. It does not require a line of sight to read the information. It can be updated with a real-time value. It requires a separate power source to read the information.

Fig. 2.13 RFID tag and reader

Working Principle

It consists of three main components. 1. scanning antenna, 2. transceiver and 3. transponder. The combination of scanning and transceiver is called RFID reader. It is a network connected device. It can be used in both portable and permanently attached forms. Readers use radio waves to transmit signals which, in turn, activate the tag. Once the tag is activated, it starts to send back the signal to antenna. It is received from the transceiver as a data through the antenna. Usually, transponder is located in the RFID tag itself. There are wide varieties of tags based on the frequency range, interference value, type of tag, and reader type. In general tag, which has larger power source, can cover longer range detection. Figure 2.13 shows the model of RFID tag and reader.

Types of RFID tags

There are three main types of RFID tags based on power sources. They are active RFID tag, passive RFID tag and semi-active RFID tag.

Active RFID tag

Active RFID systems use battery powered RFID tags that continuously broadcast their own signal. Active RFID tags are commonly used as "beacons" to accurately track the real-time location of assets or in high-speed environments such as tolling. Active tags provide a much longer read range than passive tags, but they are also much more expensive.

Passive RFID tag

Passive RFID systems use tags with no internal power source and instead are powered by the electromagnetic energy transmitted from an RFID reader. Passive RFID tags are used for applications such as access control, file tracking, race timing, supply chain management, smart labels, and more. The lower price point per tag makes employing passive RFID systems economical for many industries.

Semi-active RFID tag

The inclusion of a battery in semiactive tags allows additional features such as sensors, real-time tracking and sound notifications to be applied to the tags. The only additional feature available in onboard transmitter which cannot be semi-active tag since it's the main attribute that differentiates semi-active tags from active tags. Without an onboard transmitter, the read range of these tags is limited, and with the inclusion of a battery so is their lifespan.

The inclusion of a battery and its lack of an onboard transmitter not only differentiates semi-active from passive and active tags, but, gives them the ability to support additional features without increasing read range. Because of these features semi-active RFID is best suited for applications where additional features such as environment monitoring are necessary and the tagged items remain within range of the reader.

Active RFID tag has its own power source whereas passive RFID tags receive power from the reading antenna whose electromagnetic wave receives power from the reading antenna. In semi-active RFID tag battery runs the circuit operations while communication powered by RFID reader. RFID tags have the ability to store less than 2000 KB of data including unique identifier and serial number.

Based on the frequency range, the RFID system has three main types which are given in Table 2.9.

Table 2.9. RFID types based on frequency ranges

Frequency	Band	Range
LF RFID	30-500 kHz, typically 125 kHz	Less than three feet
HF RFID	3-30 MHZ, typically 13.56 MHz	Less than six feet
UHF RFID	300-960 MHz, typically 433 MHz	25+feet
Microwave RFID	2.45 GHz	30+feet

1. Low frequency RFID system which ranges from 30 kHz to 500 kHz with a typical range of 125 kHz. It has a short transmission ranges from a few inches to less than three feet.
2. High frequency RFID system which ranges from 3 MHz to 30 MHz with 13.56 MHz as typical frequency. It has a transmission from a few inches to several feet.
3. Ultra-high frequency RFID system ranges from 300 MHz to 960 MHz with typical 433 MHZ and can provide transmission from 25 plus feet away.
4. Microwave frequency RFID system ranges from 2.45 GHZ and can read more than 30 plus feet away.

RFID Applications

- Tracking and management
- Pet and livestock tracking
- Inventory management and asset tracking
- Cargo and supply chain logistics
- Vehicle tracking
- Employed in healthcare, manufacturing industries, retail, business care and home use.

2.4 ACTUATORS

Actuating is the process of converting energy into useful motion. Actuators are usually categorized as an energy source which requires generating some motion. There are different types of actuators available based on the application. They are:

- Pneumatic actuators use compressed air to generate motion. For example, automatic door opening mechanism. Sensors receives signal and give control to the actuator to either open or close the door.
- Hydraulic actuators use the fluid to generate some motion.
- Electric actuators use an external power source to generate motion.
- Thermal actuators use a heat source for generating the motion.

The actuators are also called transducers because they change the physical quantity from one form to another form. It is usually operated by a low voltage command signal. Based on the number of stable output produced by the actuators, it can be classified as digital or continuous actuator.

Some of the basic motors and devices used as actuators used in IoT devices are listed as follows:

1. Servomotor

The servomotor structure is shown in Figure 2.14. It is a small device which contains two-wire DC motor, potentiometer, shaft and an integrated circuit. By sending the proper coded signal the shaft can rotate to a specific angular position. When a control signal is applied to servo it applies power to the DC motor until its shaft turns to that position.

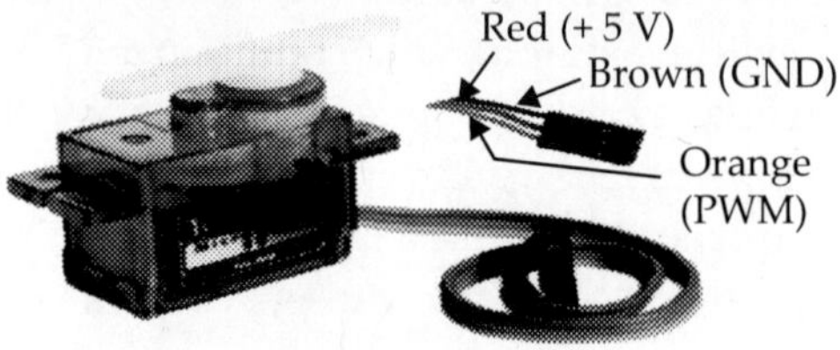

Fig. 2.14 Servomotor

(http://fabacademy.org/2020/labs/kochi/students/tom-anto/week_11(Output%20Devices)/week_11.html)

2. Stepper Motor

Stepper motors as in Figure 2.15 have multiple coils that are organized in phases. By giving energy to each phase the motor will rotate. Precise positioning and controlling of speed can be achieved by using computer controlled stepping.

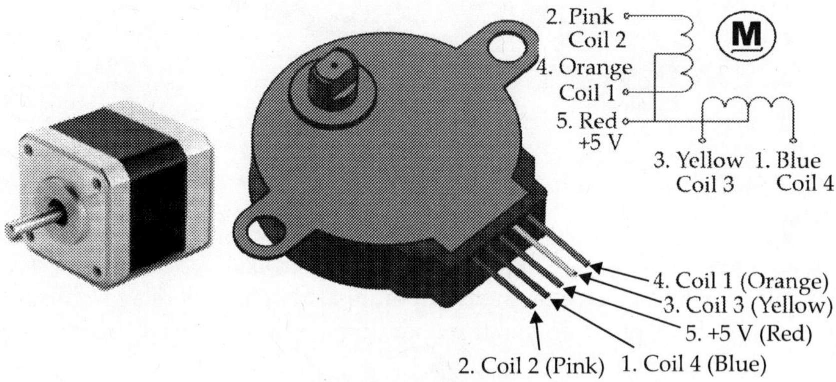

Fig. 2.15 Stepper motor

(https://components101.com/motors/28byj-48-stepper-motor)

3. DC Motor

A normal DC motor is shown in Figure 2.16 that would have only two terminals. Since these terminals are connected together only through a coil they have no polarity. Reversing the connection will only reverse the direction of the motor.

Fig. 2.16 DC motor

(https://technobyte.org/arduino-dc-motor-single-multiple-motors-interface-code/)

It is one of the most common actuators used in the IoT system. Based on the application motors of different sizes are available. It consists of permanent magnet and loops of wire inside. Whenever current flows in the motor it generates the magnetic field which reacts against the outside field of the static magnet.

4. Linear Actuator

The linear actuator device is shown in Figure 2.17. It creates the motion in a linear line. Its working is contrast to the circular motion of a conventional electric motor. It is used in disk drives, printers, dampers, etc.

Fig. 2.17 Linear actuator

5. Relay

The relay model is shown in Figure 2.18. It is a electrically operated switch. Many relays use an electromagnet to operate a switch. This relay requires only a small amount of power for operation. It can be used to control motors, heaters, lamps or AC circuits.

Fig. 2.18 Relay

6. Solenoid

A solenoid (Figure 2.19) is a specially designed electromagnet which is less expensive and is used only for on or off operations. It is frequently used in home appliances, office equipment, automobiles and pin ball machines.

Fig. 2.19 Solenoid

Application

1. Figure 2.20 gives the water level alert application for an actuator. In this the sensors sense the moisture level of the soil and send the value to the control center. Then control center commands to the water pump to either turn on or off.

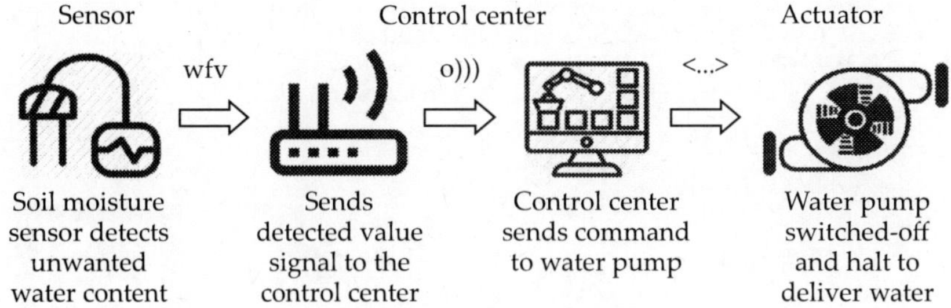

Fig. 2.20 Sensor to actuator work flow

(https://www.novatec-gmbh.de/blog/sensors-and-actuators-of-the-iot/)

**wfv –Water fraction by volume*

2. Another application for sensor to actuator flow is represented in Figure 2.21.

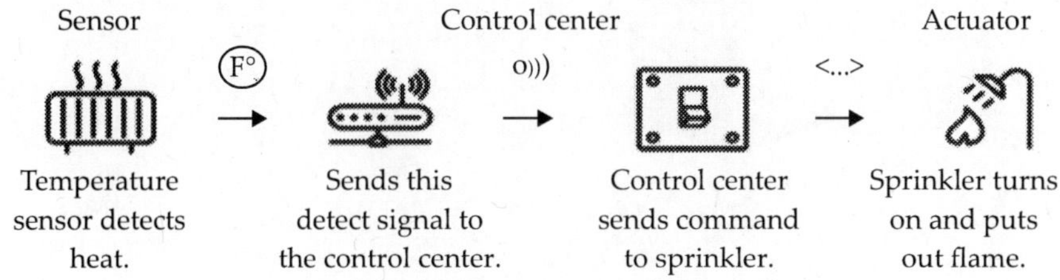

Fig. 2.21 Sensor to actuator work flow

(https://bridgera.com/sensors-and-actuators-in-iot/)

   In this temperature sensor detects the heat level and send the information to the control center, then the control center commands the sprinkler to either turn on or off.

3. Actuators are mostly used during the manufacturing and assembling process. A type of impeller cylinder uses stored compressed air energy inside a metal cylinder to move the working piston when the air is released or compressed.
4. In robotics various types of grippers are triggered by actuators with compressed air and act like human hands or fingers to perform some grabbing, drawing, pouring, granular jamming, etc.

2.5 WIRELESS SENSOR NETWORKS (WSNs)

Technological advances in wireless communications and electronics field, enabled the development of low-cost, low-power; multi-functional sensor nodes that are smaller in size and can communicate in short distances. These tiny and generally simple sensor nodes consist of sensing units, data processing, and communicating components. Figure 2.22 shows the model of IoT based WSN.

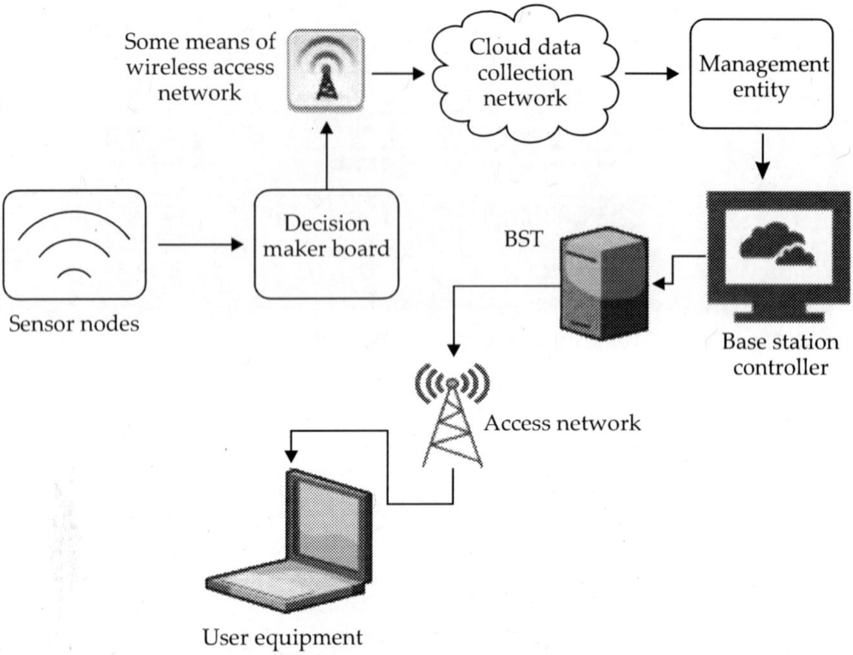

Fig. 2.22 Model of WSN

Wireless sensor networks (WSNs) consist of a number of devices that collect the information and communicate with each other through the wireless links. It consists of the base station, gateways, routers and a number of wireless sensor nodes. These sensor nodes are used to monitor the physical or environmental conditions like sound, temperature, pressure and pass the data to the main location through the network.

The data from sensors is acquired, processed and controlled by the decision making board. Then it can be transmitted to cloud through some means of access network. The management entity manages and controls the flow of processed data and then transmit it to the receiver again using some accessing network. For every smart environment the sensor networks are the key elements to gather information and it has some self-organizing capabilities based on the conditions. For these kinds of networks wide knowledge is required from a variety of applications. A few requirements are necessary to be satisfied by the sensor network. They are ease of installation, self-identification, self-diagnosis, reliability, time awareness for coordination with other nodes, some software functions and DSP, standard control protocols and network interfaces.

Open nature deployments and communication protocol developments and experiments show that WSN is one of the optimized networks for reliable operations which are time consuming and costly. It hardly satisfies the IoT applications with requirement for long-term, low-cost and reliable service, unless reusable hardware and software platforms are needed, including flexible Internet, that enable the servers to collect and process the field data for IoT applications.

Wireless sensor networks are increasingly gaining the impact in our day-to-day lives. They are finding a wide range of applications in various domains, including healthcare, assisted and enhanced living scenarios, industrial and production monitoring, control networks, and many other fields. There are many sensor manufacturers and many networks are available. Wireless sensor network comprises three important platforms: hardware, wireless connectivity, and software and challenges part.

Hardware

The main challenge in the wireless sensor network is for lesser operation and small number of sensor nodes. A large number of companies are producing small sensors, some of them are in developed stage and some of them are in developing stage. In most of the applications, WSN communicates with either LAN or WAN through a gateway. Gateway acts as the bridge between the wireless sensor network and other network. A wireless wide area network is primarily used for low power devices which are known as low power wide area network (LPWAN).

Wireless Connectivity

For a wireless data transmission and reception, several standards are available. Zigbee, Bluetooth standard are applicable for the short range communication, the long range information are done through Internet connectivity. LORA is a form of LPWAN which provides long range low power wireless connectivity for devices which are used in smart meters.

Operating Systems (OS)

The user interface which supports to interlink the hardware, process and take decision is the operating system (OS). OS of the WSN is generally less complex than available other general purpose operating system.

- Tiny OS is the first OS designed for the WSN. It is an event driven programming model. Likewise various operating systems are developed based on the wireless applications.
- Lite OS provides Unix-like abstraction and supports C language.
- Contiki OS is used as the C programming style. It is the advanced version of LoWPAN and protocol threads.
- RIOT OS is similar to Contiki.

Energy Efficiency

Maintaining efficient energy factor of the working model is the most important challenge in the WSN. There are three important issues in the WSN: (a) Increased lifespan, (b) Robustness and fault tolerance, and (c) Self configuration. WSN consumes more power when the device is in ideal state. So it is necessary to conserve power in wireless sensor nodes by switching off the nodes when it is not in use.

Characteristics of WSN

A few characteristics of sensor networks described in the WSN are helpful to design a perfect network. The characteristics are:

1. Power consumption for nodes using batteries.
2. Ability to cope with node failures.
3. Mobility of nodes is possible.
4. Heterogeneity of nodes.
5. Homogeneity of nodes.
6. Scalability.
7. Ability to withstand harsh environmental conditions.
8. Ease of use.
9. Cross-layer design.

Traditional layer approaches have three main issues:

1. It cannot share information among different layers.
2. Unable to adapt new environment change.
3. It suffers from the interference fading problems, and access conflicts.

Applications

Wireless sensors network technology and RFID technology are combined together to open up the possibilities for even more smart devices, in which a number of solutions have been proposed. For example, the solution is provided by the Intel Research Labs in the form of wireless identification sensing platform (WISP). WISP is a passive wireless sensor network with built-in light, temperature and many other sensors. Both WSN and RFID sensor networks have their own advantages but RFID sensor networks have a low range and their communication is asymmetric while WSNs have a comparatively longer range and their communication is peer-to-peer. Moreover, most of the WSNs are based on the IEEE 802.15.4 standard, which specifies the physical and MAC layer of low-rate wireless personal area networks (LR-WPANs). The technologies that enable the integration of WSN with the IoT are a hot research topic, many solutions have been proposed for that including a 6 LOWPAN standard that allows IPV6 packets to be transmitted through the networks that are computationally restricted. Also there is ROLL routing standard for end-to-end routing solutions. Some other applications for WSN are given below.

(a) **Surveillance Camera:** The area monitoring using web cameras for storing and sending data to the main database for frequent monitoring and checking from any remote place as shown in Figure 2.23.

Fig. 2.23 Area monitoring using surveillance camera

(b) **Person Healthcare Monitoring:** Figure 2.24 gives some implanted or wearable sensors on the human body for monitoring the location of the person, health condition of the person, etc.

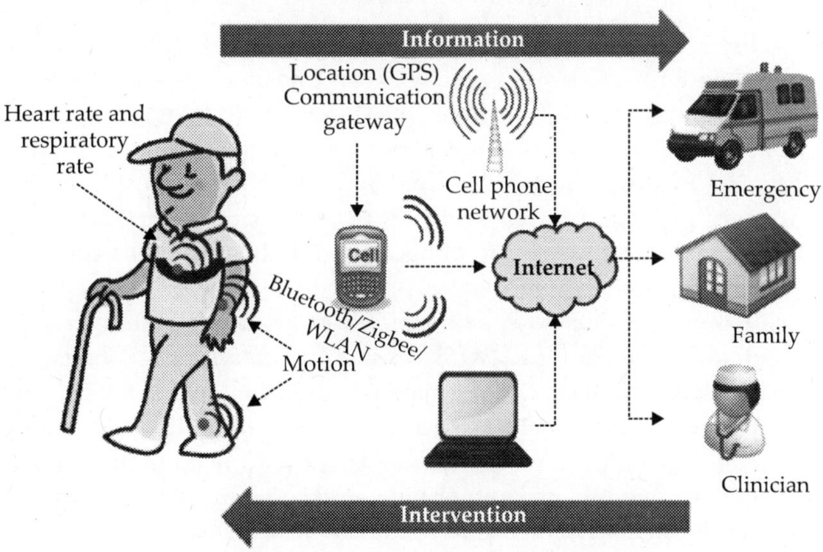

Fig. 2.24 Health monitoring using wearable sensors

(https://jneuroengrehab.biomedcentral.com/articles/10.1186/1743-0003-9-21)

(c) **Environment Monitoring:** Figure 2.25 represents general environment monitoring system which includes temperature, humidity level, etc., using ubidots IoT platform. It can also do air pollution monitoring, forest fire detection, landslide detection, water quality checking, etc., using IoT platform.

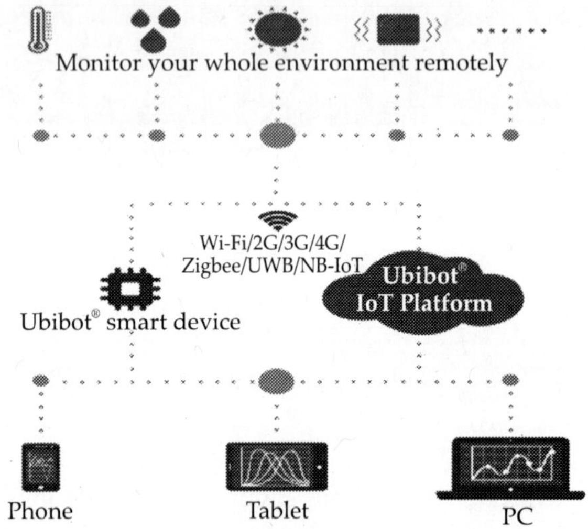

Fig. 2.25 Environment monitoring using ubidots

(https://www.ubibot.com/)

(d) **Vehicle Monitoring:** The development of smart transportation is generally led by governments or transportation authorities. Successful examples include real-time traffic monitoring and public transportation information sharing, intelligent traffic control systems, incentive program to regulate transportation, largely promotion of electric vehicle and charging facilities, and dedicated short-range communication (DSRC) enabled vehicular communication system.

Utilizing the vehicles (e.g., cars, buses, trains) along with the roads and the rails equipped with sensors, actuators and processing power, important traffic information could be offered to the drivers or passengers of the vehicles to achieve better navigation and safety. Benefiting from the more accurate traffic information about road patterns, governmental authorities could better plan and design the roads. Particularly, intelligent roads can be designed, with warning messages based on climate conditions and unexpected events (e.g., accidents or traffic jams).

2.6 FEATURES OF SMART DEVICES

Smart device designs are challenging because of changes day by day with booming technology and market requirements. Usually, changes occur with size, design material, application tools, add-on features, adaptable with new generation and user requirements. The features of smart devices can be split into two types. One is device level feature and another one is operation level features. Device level features are about the function, access, usage and control of the device, along with its operation. Processing or operational level feature describes the characteristics necessary at operative level.

2.6.1 Device Level Feature

Remote Access

The most basic feature of the smart devices is to control and operate the devices using remote control by RF (radio frequency) range as in Figure 2.26. Various software programs recently developed make the program sequence of the device activities and to capture the device events executed by the power line controllers. Controlling of devices is also done by working remotely. For example, enhance the ability of the disabled person to control the devices from distance itself. The basic routine work of the people gets automated by these kinds of remote devices which showcase the advancement and success of the technology.

Fig. 2.26 Remote controlled devices

(https://www.gearbest.com/blog/how-to/3-main-types-of-remote-control-and-their-applications-2769)

Device-to-device communication

In device-to-device (D2D) communication, the devices can communicate directly with each other without using any communication infrastructures like access point (AP) and base station (BS). With the advancement in technology various new trends in the applications give rise increase in standards of the ordinary devices to smart device. Let us discuss what is device and how it is necessary to communicate with device-to-device in detail.

What is a device?

The device is an electronic based system which consists of sensors, actuators, processors and logic gates. These devices can be used as input devices like sensor, relays, and output devices like smart phone, any other smart electronic display and controlling devices like actuators, relays, and switches.

Device-to-device

If both the input and output devices are sharing the information it is known as a device-to-device communication. For example, consider the air conditioner. The control action is transfer from the smart phone to the main unit of the air conditioner machine to switch ON or OFF, to increase or decrease the temperature by adjusting timer, etc., as shown in Figure 2.27.

Fig. 2.27 D2D communication

The system based devices have the capability to transmit and to receive the information from anywhere in the world over the Internet communication. This kind of communication is referred to as Internet of Things. This way of connected environment is the main focus of many industries for remote monitoring and automation. With these features anyone can get the information of weather forecast and can predict the climate before taking any action. There are two tiers of D2D communication available. They are device tier and macro cell tier. In the device tier, two devices can independently communicate and share the information with each other without any base station node as shown in Figure 2.28 (a). In macrocell tier, the device-to-device communication can be made by using base station as given in Figure 2.28 (b).

For device-to-device communication, the device requires location discovery and communication to the neighboring devices with the availability of functions, which is to reduce the cost of communication among the devices. This device-to-device communication plays an important role in mobile Cloud computing for effective sharing of the

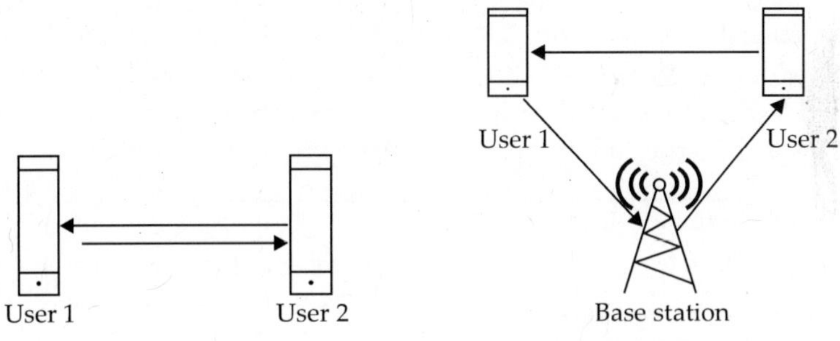

Fig. 2.28 Two tiers of D2D model

resources like spectrum, computational power, computational time, energy, social contents and various other applications. The important advantage of this device tier communication is that, it can be developed as temporary communication between D2D within a short period when any unexpected or uncertain Internet communication or network infrastructure failure occurs due to natural disasters like earthquake, forest fire, building collapse, etc., for rescue purpose.

The infrastructure-less connection trend started to increase with the current technology for achieving fast response and for quick action. At the same time, the D2D model gives more benefits, but suffers from interference problem due to the multi user with single device by the utilization of unlicensed band. In addition to this it also faces challenges in security issues, resource allocation problem and the quality of service. Consider any particular device which, when tethered to the Bluetooth speaker, that device alone functions best with the speaker. If more than one device is connected to the speaker it turns to cause the interference. Nowadays, the researchers have started to analyze how to improve the efficiency of resource allocation, how to provide secured D2D communication and how to manage the interference issues in cellular system.

People can become smarter in their day-to-day life by casting their mobile with smart TV and IoT devices. Most of our regular activities are monitored by the mobile itself with its Apps as shown in Figure 2.29 like health monitoring. It monitors pressure level, step tracking, heart rate monitoring, remote medical assistance, etc. In case of vehicle safety, it provides theft alert, fuel alert, driver attention monitoring (sleeping, drunken and drive) and the parking allotment. In smart home, it automates the light, fan, AC control functions and smart electricity meters are used as device-to-device communication. Activation of one device will-activate other devices in sequence is the main advantage of D2D and it leads to the more sophisticated environmental lifestyle. Consider, for example, if the doorbell rings, then the TV sound gets muted automatically and displays the face of the guest who is waiting outside or, mobile phone call forwarded to the office landline when the user stays in smart environment and not picked up the call in mobile phone in emergency condition.

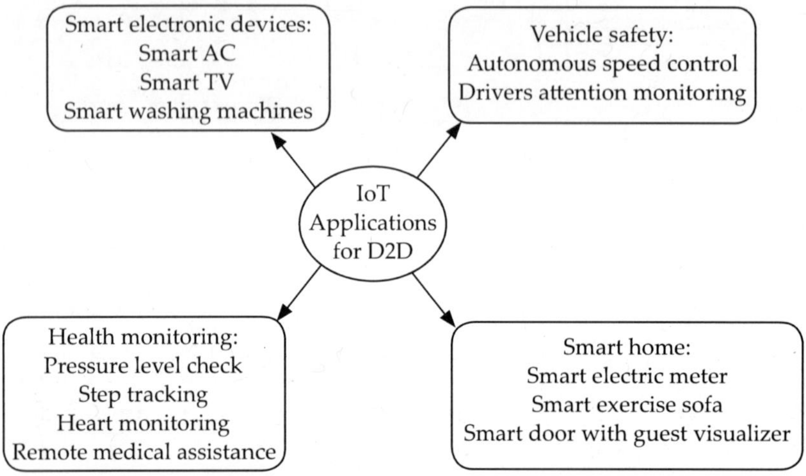

Fig. 2.29 IoT applications in D2D communication

2.6.2 Processing or Operating Level Feature

Another feature of the smart devices is based on its operating level. In this, the signals are acquired, processed, stored and then taking decision according to the conditions specified in the program as per user information. There are four important features in operating level which are discussed as below:

Information Acquisition / Dissemination from intelligent sensor networks

Development of sensor based technology tremendously increases in sharing information in low level decisions. As a result of information acquisition, the environment can be modified based on the sensor readings.

Figure 2.30 shows the flow of signal processing and data transmission. In this process, the captured data from the sensor are passed to the data acquisition system for data aggregation and for pre-processing then it transmits the processed information to the Cloud database via middleware network. For example, consider the smart exercise sofa which measures an individual weight and modifies the settings in the device according to their status to sit comfortable by varying the cushion level. The presence of individuals in the surroundings can be detected using some motion detector sensor and pressure sensor. Based on these, lights and music players in the house can be controlled accordingly. Usage of gas sensor and water sensors will monitor the leakages in the gas valve and the force valve to close when danger arises.

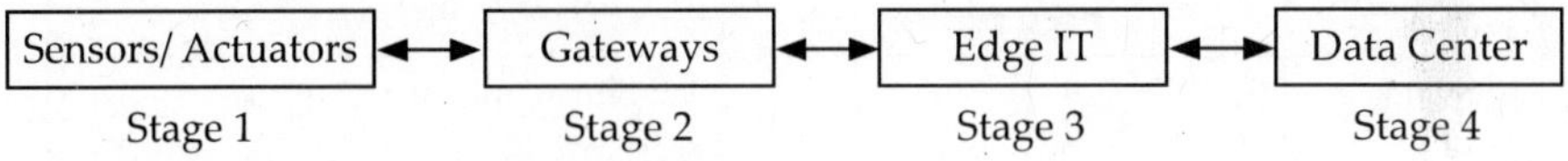

Fig. 2.30 Information acquisition/dissemination

Enhanced Services by Intelligent Devices

Smart devices are usually fitted in the smart environment monitoring system. The impact of these devices becomes more powerful when these devices are attached to the intelligent sensor. For example, the remote care givers can continuously monitor their patient's status using smart environment. Various manufacturing companies are frequently monitoring the status of the product parts in production line from the suppliers. Researchers are currently working on the concept of converting time consuming manual work into automatic work. A standard company Philips invented a kind of tablecloth which charges all the chargeable equipment placed on it. An environment that can combine with the features of these devices, the information gathering can remotely control the power based on the application need in an automatic manner.

Predictive and Decision Making Capabilities

The features of a smart environmental system gives the potential for fulfilling the goal of a smart environment to improve the experience of inhabitants of the environment. However, controlling of these capabilities is mostly done by the hands of the user. Only through explicit remote manipulation or careful programming of these devices, sensors and controllers can adjust the environment to fit the needs of the inhabitants. Behavior of inhabitant as well as the environment can be predicted based on the observed activities and the known features. For example, an intelligent car smart system can collect information about the driver's behavior, timing of travel and optimal routes to destination details and it displays the information to the driver.

Networking Standards and Regulations

A smart environment will be able to control and to manage all of its various networked devices from anywhere and at any time through the Internet. All of the hardware and software should be operated based on certain standards and regulations. Moreover, it is easy to install, configure and operate in order to be user friendly to all the consumers. The various wireless standards and their purpose are listed in Table 2.10.

Table 2.10. IoT Networking standards and regulations

Standards	Frequency	Purpose
IEEE802.11	2.4 GHz	Wi-Fi for medium range transmission.
IEEE802.15	2.4 GHz	Zigbee
Bluetooth	2.4 GHz	Short range transmission
FHSS	2.4 GHz	Modulation technique
OFDM	2.4 GHz	Modulation techniques
Wired:		
Ethernet	Up to 500 MHz	For fastest data transmission
X10 Power line standard	310 MHz in the U.S. and 433.92 MHz in European systems	X10 is a communication protocol enabling communication between smart home devices.

2.7 EMBEDDED DEVICES IN SMART PRODUCT

The devices which are smart to interface and communicate can be controlled by Internet is called IoT devices. Smart devices are usually in-built with the embedded hardware and software. The embedded hardware part comprises various components in it. The components include sensors, actuators, antennae, memory chips, power supply cable ports, etc. Each component in the embedded unit has some unique purpose. Based on the specified smart operation, IoT devices are programmed accordingly (Figure 2.31).

Architecture is the internal structure of the embedded system. Structure in Figure 2.31 consists of a number of input and output units, processing and control unit, and communication unit.

I/O Unit

The I/O unit consists of input device, output device and notification part. Different types of sensors, switches, timers are the input devices to get input from the application environment and sends to the processing unit. After processing, the output can be displayed in LCD / seven-segment display. Meanwhile the notification can be sent to the mobile via SMS / Email.

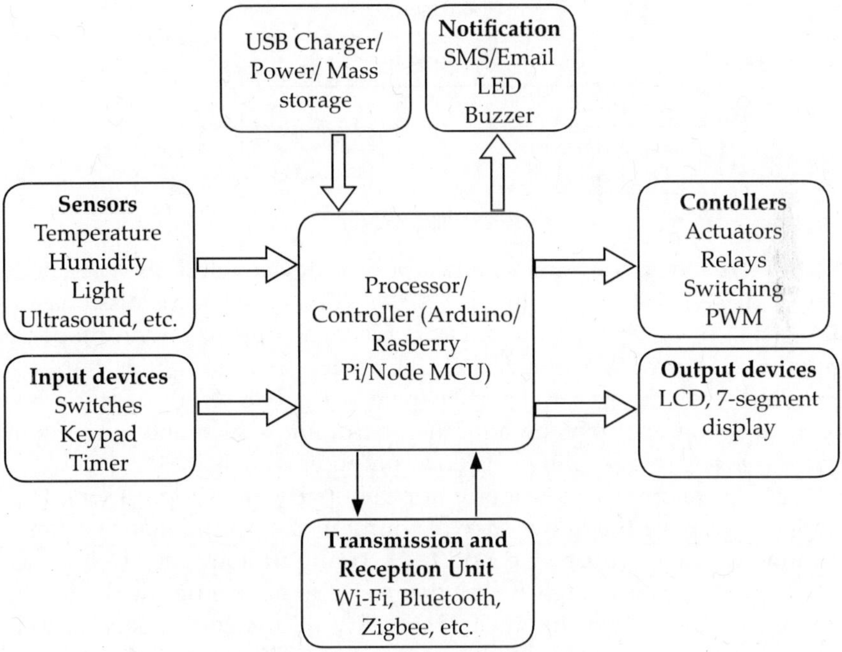

Fig. 2.31 Architecture of IoT embedded setup

Processing Unit

It comprises both processing and controlling part. Based on the input data from the input device the processor will analyze the data and provide control to the device using some controlling components. The controllers used here for this purpose are actuators, switches, relays and PWM.

Communication Unit

It is a transreceiving unit consisting both of transmitter and receiver. The processed information from the processor can be transmitted to the user via some long and short range wireless / wired devices. For wireless access, Wi-Fi, Bluetooth, Zigbee can be used. For wired access USB cable and ethernet cable can be used. For Wi-Fi and BT modules, separate chipboards integrated with board of the controller or processor in-built boards can be used. For example, the Wi-Fi module ESP8266, ESP32, ESP12 along with Arduino can be used. Alternatively Node MCU, Rasberry Pi with inbuilt Wi-Fi module can be used.

For IoT implementation the power supply to the board can be provided by USB charger or separate power supplies. USB ports comprise USB-Serial chips, USB-Serial port, and USB-Serial output header as shown in Figure 2.32.

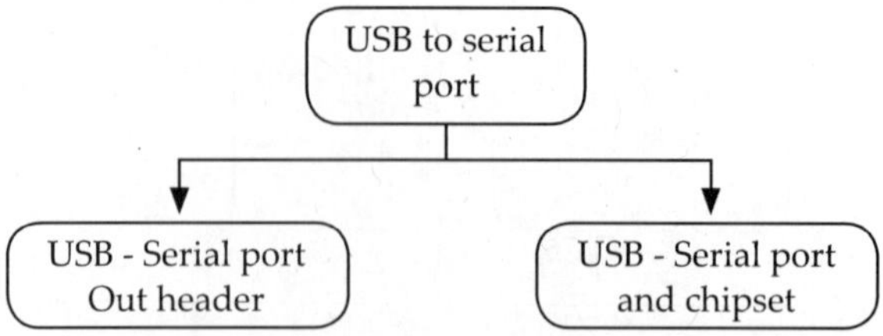

Fig. 2.32 USB serial port

USB-Serial chip develops and supports devices and their related software drivers for converting RS-232 or TTL serial transmissions to USB signals, in order to allow support for legacy devices with modern computers. If need to connect a serial device to a computer but the computer doesn't have a serial port, then need a USB to serial adapter. USB headers are groups of pins on a motherboard, into which one can plug a special cable to provide USB ports on a computer chassis.

While developing any device for the particular application, the designer must have the knowledge of sound in the component selection, and how to make proper connection, configuration and write the appropriate program to activate the device. After activating the device, it is necessary to maintain the device reliability till the end of service and analyze the challenges while utilizing the device. Figure 2.33 represents the motherboard model for IoT embedded system.

The embedded system can read the real-time data from the sensors. After getting the data from input devices, the system processes the received data and then stores information in its internal memory or memory storage

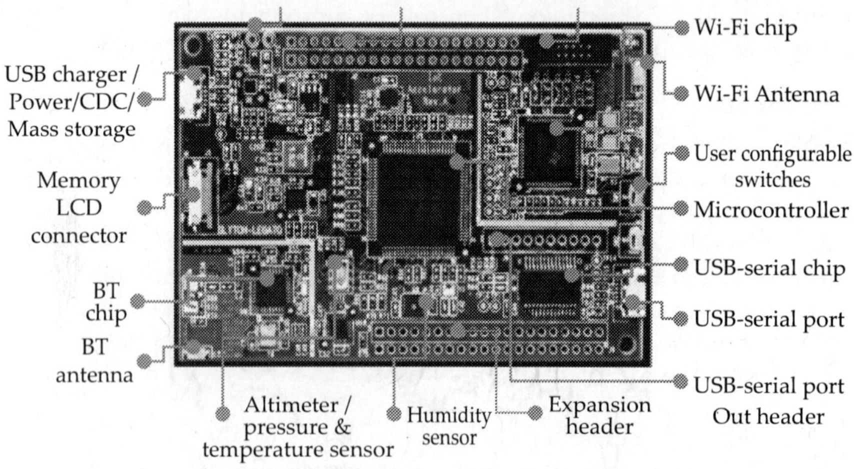

Fig. 2.33 General motherboard structure of IoT embedded system

(https://images.app.goo.gl/Ng519ac7r8frbnfS7)

devices like SD cards. It can also receive the input from the local network which consists of multiple users. This local network supports to connect the user from any kind of local area network (LAN), wide area network (WAN), metropolitan area network (MAN) and MANET.

The processed data and the stored data will be transmitted to other devices through the Internet. The data may be in any form of text, voice, image and video. Based on the input and the processed data, the system will take the decision based on the program and transmit the control information. For example, if anyone touch or try to open a car in the parking, owner will get information in their mobile or any smart gadgets. In the existing design it just raises the alarm when any one touches the car. Nowadays, it is possible to capture the image of the person and send it to your mobile. Another example is automatic door bell decision making system. This is one of the interesting and secured concept for the home alone people. Instead of immediately opening the door after ringing, one can view who has rang the bell in your smart TV or mobile or some gadget monitor.

2.8 CHALLENGES IN DESIGNING IoT SYSTEM

Design of embedded system for a particular task has compressed more issues. After taking it into the account of any application development, the following criteria have to be considered.

(a) have to collect the components related to the specific design like processor, sensor, and cables called the embedded devices;
(b) to get the proper output;
(c) to provide the long life or reliable service;
(d) to select the life-time of the device;
(e) to operate in secured way; and
(f) to provide the flexibility.

These are design parameters one has to follow in the design. After designing, or releasing the product to the market, it is possible to face some issues in the devices and consider it a challenge. There are six important challenges usually faced while designing the device. They are explained below:

Flexibility of Various Apps over Designed System

The upcoming devices should be adaptive to the heterogonous network. It should support various new applications. For example, consider mobile with Android software with various supporting applications running on it. Due to this kind of multifunction usage in the phone, they usually suffer from the major issue of charge reduction and power dissipation problem. In general, problems faced by the IoT devices are:

1. Integrity between various services.
2. Difficult to run in new environment.
3. Fluctuation due to adding and deleting of Apps.
4. Space occupation and power value of components.
5. Energy consumption factor.

Various researches are currently going on to solve the above problems and trying to suggest proper solution for these challenges.

Security Issue

It is necessary to ensure that the communication has been secured between the devices from hacking. The intruder (stranger) can easily access the information without any knowledge of concerned person. Various new encryption algorithms are being developed day by day to enhance the security level.

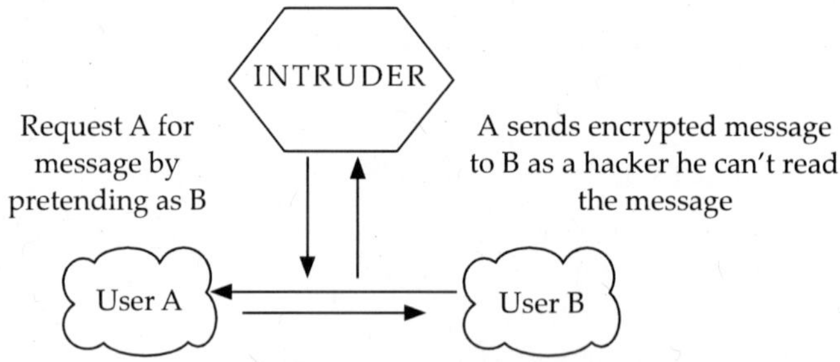

Fig. 2.34 Hacking of transmitted data

Suppose two users A and B want to share their information between them without any leakage and loss in a more secured way, and if the intruder hacks the message means it suffers from the loss of security as shown in Figure 2.34. But if proper encryption algorithm prior to the transmission is selected means the information they received as an encrypted cipher text message like (@3$5) to the hacker which was very tough to decrypt and the original information can be transferred in secured manner.

Power Management

The prime factor among all other challenges is power consumption. Nowadays, all digital devices are handheld and attractive in size, shape for usage. Always carrying the device for frequent updates leads to think about the device life factor. It is very important to design low power consumption devices with long reliability. Nowadays, the device travels like a human. The digital computation technique replaces most of the conventional analog devices. Simply we can say, most of the machine tasks are reduced by computerized monitoring. In future it may replace the human also. Whenever the human felt weak or hunger could have to take some food to get energy. Likewise, if the device's battery power is low, it is necessary to recharge it to regain its power. The main challenge is how to quickly boost up its energy with less power reduction. For this kind of power boosting, selective transistors are used in the devices which should be designed for consuming less power. The gate size of the transistor should be kept small for low power consumption value. The designer should have concentrated more and plan the device size as such for less power value and the biggest challenge for the effective power management.

Verification and Testing

Verification and testing of IoT product is similar to general equipment testing. It has been done in two levels: product level and operation level. The product level testing is carried during the product developmental stage. After manufacturing, the device has to be completely tested including for operation level testing during the product operational stage. This will absolutely explore its basic operation behavior. It is essential to approve the device which shall meet all the requirements and also provide the assurance about the quality, life-time, i.e., warranty and guarantee.

Safety Measures

It is essential to provide the functionality safety of the device for proper working condition and need to take a number of safety trials for its verification. In case of any unconditional event happen then the alternative solution should be in-built with the device itself as a precaution. For example, in hardware level, open circuit or isolation of the device by fuse can be used in case of any emergency. In software level, shutting down of the application or suspend the task can be used.

Delivery Time

Timely delivery of product to the market is an important target for the designer. Meanwhile cost of the product is another major challenge. The product should be made cost effective to the consumer standards and deliver to the market. Many of the industries are doing this nowadays and are appreciated from the customer for timely delivery without any delay. For example, the leading industries having the prescribed target time for manufacturing the standard product with good delivery in time as their policy to lead a successful industry.

2.9 SUMMARY

In this chapter a detailed discussion of IoT devices and their pin diagrams with their specific application has been done. The characteristics of various IoT sensors and their functions discussed. Wireless sensor network used for developing IoT applications along with how to design smart system and also how to convert the conventional system into smart system.

REVIEW QUESTIONS

1. Write the working principle of sensors.
2. What are the various types of sensors used in IoT with examples?
3. Compare the different types of RFID.
4. Define actuators and their types.
5. Describe some basic motor models used in actuators.
6. List the components in the architecture of WSN.
7. What is WSN? List some characteristics of WSN.
8. Explain any two applications of WSN.
9. What is D2D communication?
10. Define 2-tier and macro cell tier of D2D communication.
11. Write down the features of smart devices.
12. Design the architecture of IoT embedded setup for any specific application.
13. Describe the IoT designing challenges faced during for specific application.

3

TECHNIQUES FOR IoT

3.1 INTRODUCTION

Internet of Things (IoT) is a unified network of physical objects that can change the parameters of the environment of their own. It has the capability to gather information from the environment and transmit it to other devices. It is an emerging field in the development of the Internet. This technology will give immediate access to the information from the physical world and the objects in it, leading to innovate the services and increase in efficiency and productivity. The IoT is enabled by the latest developments, smart sensors, communication technologies and the Internet protocols. By the way of enhancing latest developments, communication over power line which emerges with a good start of sending data over the power lines. It is explained briefly along with its characteristics in this chapter.

3.2 CONVENTIONAL POWER LINE NETWORK

General power line carries the electric power from power distribution system and is delivered to the individual customer through the power line. Distribution substation carries very high voltage which can be converted into medium voltage by the transformers in the range of 2 kV and 5 kV. Then the primary distribution lines carry this voltage and reduce further while transmitting to the premises (customer buildings). Then the reduced voltages are carried by the secondary distribution line. The consumers are connected to the secondary distribution lines through service drops. Consumers who are in need of high voltage can directly connect to the primary distribution line. Usually, for home purpose 220 V to 240 V, 50 Hz (5 A to 15 A) AC supply is applied. Alternating current (AC) mainly depends on the frequency value. By using this frequency concept, the idea of power line communication e merges. In a normal power distribution, four outlet wires are used. Each wire out of three wires used for three-phase supply remaining one wire is for ground purpose.

The conventional power line consists of generation station, substation, transmission lines and sub-transmission lines as shown in Figure 3.1. The generation station is the main source of the power station where power is generated. This power is transferred to the transmission

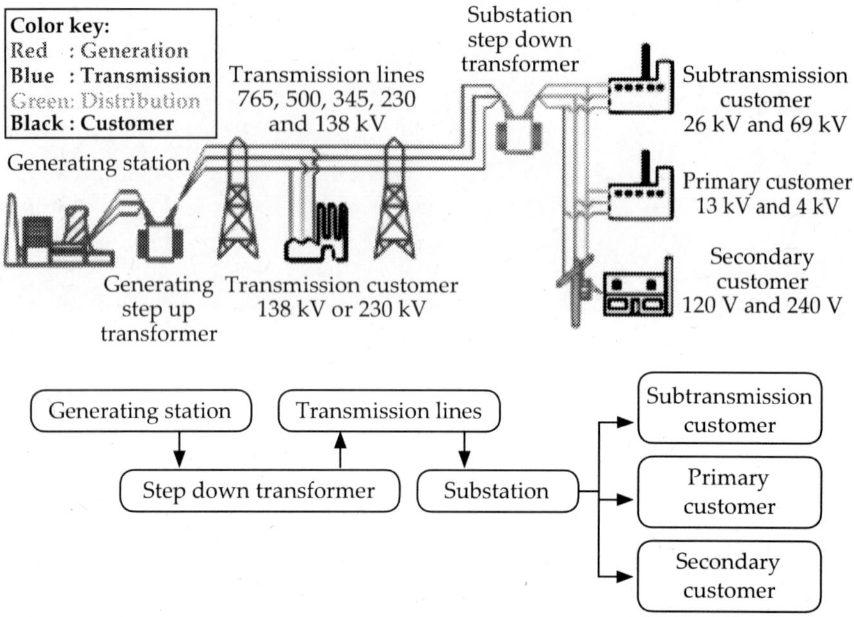

Fig. 3.1 Conventional power line

(https://en.wikipedia.org/wiki/Electric_power_transmission)

line with different voltages like 138, 230, 345, 500 and 765 kV through the stepup transformer. The high voltage on transmission lines are delivered to the customer as low voltages via sub-transmission lines like 24 kV, 120 kV of different voltages through the step-down transformer. For a power line communication, data is sent through the power lines by superposing a modulated high frequency carrier signal to the high/ medium/low AC or DC supply line. The brief concept of power line communication is described in the following section.

3.2.1 Power Line Communication in IoT

Power line communication (PLC) is an important and essential part of the IoT because power lines are everywhere. PLC emulates the communication technology to send data over the existing power cables. This means that the electrical power cables run to an electronic device to carry the information. For example, one can do both power it up and at the same time control/ retrieve data from it in a half duplex manner. PLC is also called power line carrier, power line digital subscriber line (PDSL), mains communication, power line telecommunications, or power line networking (PLN).

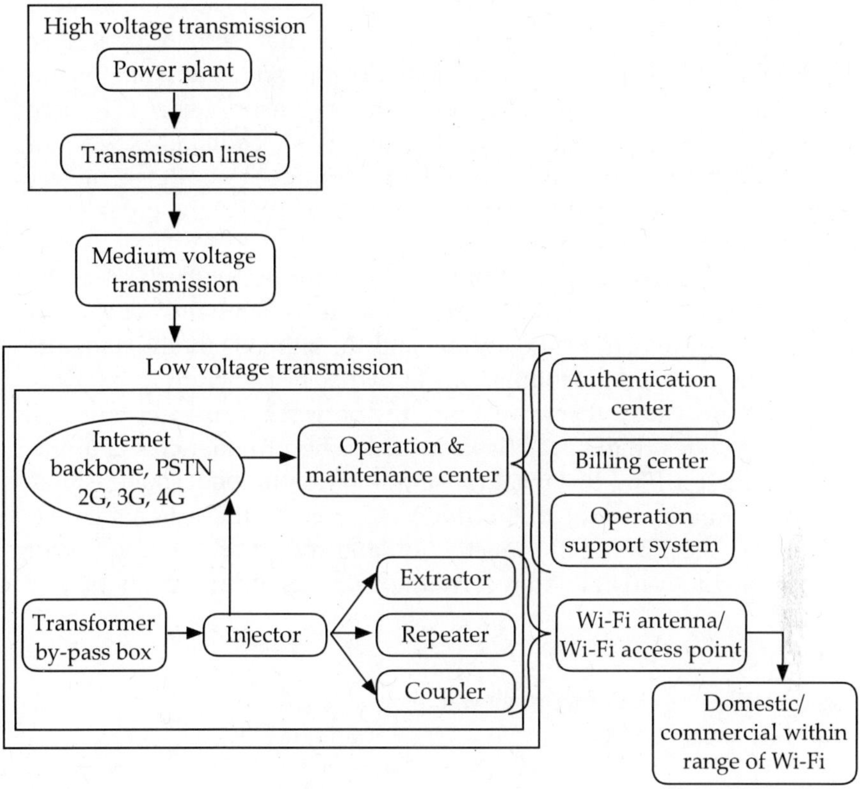

Fig. 3.2 BPL transmission and reception

(https://electronicsforu.com/technology-trends/broadband-power-lines)

Power line communication systems are operated by adding a modulated carrier signal to the existing power wiring system. A wide range of power line communication technologies are needed for different applications such as home automation to the Internet access which is often called broadband over power lines (BPLs). The BPL communication is shown in Figure 3.2. The conventional power line units like power plant, transmission lines, substation and customer substation are considered IoT network, operator and maintenance center and operating support system. Most PLC technologies limit themselves to one type of wire (such as premises wiring), but some can cross between two levels (for example, both the distribution network and premises wiring). Typically, transformers prevent propagating the signal, which requires multiple technologies to form very large networks. Various data rates and frequencies are used in different situations.

3.2.2 Working of BPL

In order to inject data signal into the power signal, certain broadband components can be added to the transmission line. The components are Internet injector, coupler, repeater and extractor. Injector is used to provide interface between power line and the broadband power line system. It aggregates the user data in the power line. The BPL signals are injected into and extracted from medium voltage (MV)/low voltage (LV) lines through inductive or capacitive couplers.

In an inductive coupling, BPL signal couples onto the line by wrapping inductors around the line. On the other hand, capacitive coupling uses a capacitor for coupling and the signal is modulated onto the line voltage. Inductive couplers are known to be rather lossy, but since they require no physical connection to the network. It is safer to install on energized lines than capacitive couplers. High frequency signal line is not suitable for long distance transmission due to the frequent signal loss and attenuation problem. In order to overcome these signal issues, repeaters are used between transmission and reception lines. Extractor in BPL is used to transfer and receive the data signal to consumers.

3.3 TYPES OF PLC

In general, PLC can be broadly viewed as:

1. Narrowband PLC
2. Broadband PLC

3.3.1 Narrowband PLC

Narrowband PLC works at lower frequencies from 3 to 500 kHz with lower data rates up to 100 kbps, and has longer range up to several kilometers, which can be extended using repeaters. Broadband PLC works at higher frequencies—ranges from 1.8-250 MHz—with high data rates up to 100 Mbps and is used in short-range applications. Narrowband power line communication is receiving widespread attention due to its applications in the mart Grid. Another application that narrowband PLC has been used is smart energy generation, particularly in microinverters for solar panels.

Narrowband power line communications began after electrical power supply became widespread. Around the year 1922 the first carrier frequency systems began to operate over high-tension lines with frequencies of 15 to 500 kHz for telemetry purposes. In the 1930s, ripple carrier signaling was introduced on the medium (10-20 kV) and low voltage (240/415 V) distribution systems. In the 1970s, the Tokyo Electric Power Co. ran experiments which reported successful bidirectional operation with several hundred units.

Since the mid-1980s, there has been a surge of interest in using the potential of digital communication techniques and digital signal processing. The drive is to produce a reliable system which is economical enough to be widely installed and able to compete cost effectively with wireless solutions. But the narrowband power line communications channel presents many technical challenges, a mathematical channel model and a survey of work is available. One natural application of Narrowband power line communication is the control and telemetry of electrical equipment such as meters, switches, heaters and domestic appliances.

Control and telemetry applications include both utility side applications, which involve equipment belonging to the utility company up to the domestic meter, and consumer side applications which involve equipment in the consumer's premises. Possible utility side applications include automatic meter reading (AMR), dynamic tariff control, load management, load profile recording, credit control, pre-payment, remote connection, fraud detection and network management, and could be extended to include gas and water.

Open smart grid protocol (OSGP) is one of the most proven narrowband PLC technologies and protocols for smart metering. There are more than five million smart meters, based on OSGP and using BPSK (binary phase shift keying) PLC, installed and operating around the world. The OSGP Alliance, a non-profit association originally established as ESNA in 2006, led an effort to establish a family of specifications published by the European Telecommunications Standards Institute (ETSI) used in conjunction with the ISO/IEC 14908 control networking standard for smart grid applications. OSGP is optimized to provide reliable and efficient delivery of command and control information for smart meters, direct load control modules, solar panels, gateways, and other smart grid devices. OSGP follows a modern, structured approach based on the OSI protocol model to meet the evolving challenges of the smart grid.

3.3.2 Broadband PLC

Broadband PLC has become one of the most important communication technologies for smart gird applications, especially for AMI (advanced metering infrastructure) and HAN (home area network). Currently, most broadband PLC products are based on OFDM, DMT or spread spectrum technologies. Their performances differ from each other on certain circumstances. Broadband PLC technology transmit high speed Internet through the accessible electrical paths and it enables the data transmission between 500 kbps and 3 Mbps which is certainly equivalent and faster to most data transmission technologies. The prevalence of smart phones and high speed broadband has remarkable impact on the adoption of smart homes across the globe. Therefore, these chipsets are significant traction in smart homes owing to its high-speed connectivity.

Currently, there is an ample opportunity in developing countries where the broadband penetration is relatively low when compared to developed economies, especially in the rural areas. This technology opens an emerging opportunity for providing broadband over power lines at cheaper prices. Besides, low maintenance costs and lesser installation time, it makes broadband PLC, a feasible technology for increasing broadband penetration in developing economies. Broadband over power line (BPL) is a system to transmit two-way data over existing AC MV (medium voltage) electrical distribution wiring, between transformers, and AC LV (low voltage) wiring between transformer and customer outlets (typically 110 to 240 V). This avoids the expense of a dedicated network of wires for data communication, and the expense of maintaining a dedicated network of antennas, radios and routers in wireless network.

Applications

Smart grid, smart networking, lighting automation, security and surveillance, long haul communication.

3.3.3 Broadband over Telephone Lines

Broadbands over the telephone network are in four different forms. They are Internet transmission via digital subscriber line (DSL), fiber optics cable, coaxial cable and satellite.

3.3.3.1 Internet over DSL

The Data Subscribers Lines (DSLs) transmit the modulated signal over the unused telephone wires. Figure 3.3 shows how the telephone line used to carry the digital signal. Normally, telephone lines are used for carrying voice signal. Computers transmit the data signals, or binary signals. This method does not cause any interference to the already existing telephone communication services. The speed of the communication varied based on the distance of transmission line.

3.3.3.2 Coaxial Cable

The broadband cable connection is provided by the local cable TV provider. Internet access through cable connection can be represented as cable Internet as shown in Figure 3.4. It is a kind of Internet access which uses the same infrastructure as that of TV operator. For an

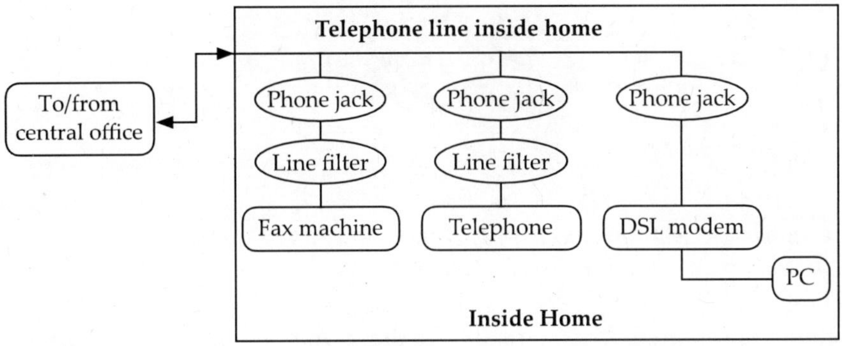

Fig. 3.3 Broadband over DSL

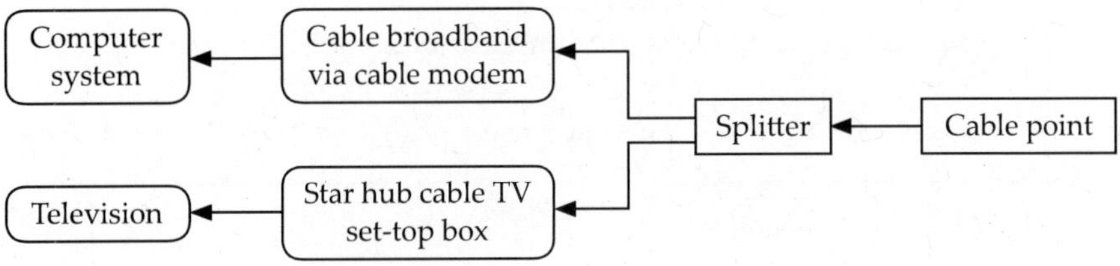

Fig. 3.4 Broadband over coaxial cable

Internet connection via cable TV line cable modem was required at the customer's side and the cable modem termination was required at the cable television head end. The modems can be connected via coaxial cable or a hybrid fibre coaxial plant. Coaxial cable or hybrid fibre coaxial (HFC) plant. Cable modem can effectively work up to 160 km. In case of HFC network for an effective management modem termination system can be grouped into hubs. Here the cable Internet connection speed varies with the number of users on the service at a specific point in time. Given a specific geographical area, users of the broadband cable service share the connection bandwidth which slows the speed when more users are on the system. This will occur at the peak times, for example, late in the evenings after the working hours since many people will be accessing the Internet.

3.3.3.3 Fiber Optic

The latest broadband service is fiber optic (Figure 3.5) which is the fastest Internet connection so far compared to other connections. However, this type of Internet service is still in its infancy as its service areas are quite limited and because the laying down of the fiber optic cable takes a while to complete. Wherever it is available, the cost not only competes with that of DSL and cable, but it provides a much faster connection than both of these services.

Fig. 3.5 Broadband over fiber optic

https://nwnc.co.uk/wp-content/uploads/2018/10/FTTP-Fibre-Broadband-Kent.jpg

3.3.3.4 *Satellite*

Satellite represented in Figure 3.6 is the common broadband service worldwide. Although this is a good replacement for dial-up for those people living in remote rural areas, the installation costs are quite high, but the ongoing monthly charges are competitive to both cable and DSL.

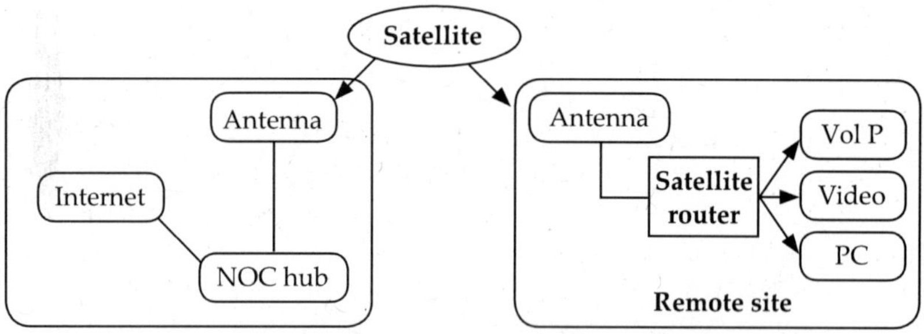

Fig. 3.6 Broadband over satellite

3.3.4. Issues in PLC

Different types of power line communications use different frequency bands. Since the power distribution system was originally intended for transmission of AC power at typical frequencies of either 50 Hz or 60 Hz. Power wire circuits have only a limited ability to carry higher frequencies. The propagation problem is a limiting factor for each type of power line communications. The main issue determining the frequencies of power line communication is laws to limit interference with radio services. Many nations regulate unshielded wired emissions as if they were radio transmitters. These jurisdictions usually require unlicenced uses to be below 500 kHz or in unlicensed radio bands. Some jurisdictions (such as the EU), regulate wire line transmissions further. The US is a notable exception, permitting limited power wide band signals to be injected into unshielded wiring, as long as the wiring is not designed to propagate radio waves in free space.

Data rates and distance limits vary widely over many power line communication standards. Low frequency (about 100–200 kHz) carriers impressed on high voltage transmission lines may carry one or two analog voice circuits, or telemetry and control circuits with an equivalent data rate of a few hundred bits per second. However, these circuits may be many miles long. Higher data rates generally imply shorter ranges. A local area network (LAN) operating at millions of bits per second (mbps) may only cover one floor of an office building, but eliminates the need for installation of dedicated network cabling.

3.3.5 PLC over AC and DC Lines

Depending upon the underlying power line characteristic, PLC can be further classified as

1. PLC over AC lines
2. PLC over DC lines

Many utilities around the world have chosen narrowband PLC over AC lines for their smart grid projects. Example: by monitoring electricity usage based on time of day and even by device or application, utility companies can provide pricing structures that give consumers incentives to adjust their energy consumption, thus reducing peak load and avoiding the need to construct new power plants.

The popularity of PLC adoption in smart grid applications has led to significant focus on PLC over AC power lines. However, narrowband PLC over DC lines is also gaining ground in home networking, lighting and solar applications as well as in transportation vehicles (electronic controls in airplanes, automobiles and trains). The use of PLC in these applications reduces wiring complexity, weight, and ultimately cost of communications.

3.3.6 Applications

3.3.6.1 Home Automation

Power line communications technology can be used with electrical power wiring within a home for home automation. For example, one can use remote control for lighting and some home appliances without installation of any additional control wiring.

Typically, home control power line communication devices operate by modulating in a carrier wave of between 20 and 200 kHz into the household wiring at the transmitter. The carrier is modulated by digital signals. Each receiver in the system has an address and can be individually commanded by the signals transmitted over the household wiring and decoded at the receiver. These devices may be either plugged into regular power outlets, or permanently wired in place. Since the carrier signal may propagate to nearby homes (or apartments) on the same distribution system, these control schemes have a house address that designates the owner.

3.3.6.2 Home Networking (LAN)

Power line communications can also be used in a home to interconnect home computers and peripherals, and home entertainment devices that have an Ethernet port. Power line adapter sets plug into power outlets and establishes an Ethernet connection using the existing electrical wiring in the home (power strips with filtering may absorb the power line signal). This allows devices to share data without the inconvenience of running dedicated network cables.

The most widely deployed power line networking standard is from the HomePlug Powerline Alliance. HomePlug AV is the most current of the HomePlug specifications and was adopted by the IEEE 1901 group as a baseline technology for their standard, published on December 30th in the year 2010.

3.4 POWER LINE CHANNEL CHARACTERISTICS

Power lines were originally designed for transmission of power at the range of 50 to 60 Hz. At high frequencies the power line is very hostile to signal propagation. A brief overview of the three important PLC channel characteristics is as follows.

3.4.1 Attenuation

In general, attenuation is a kind of loss which reduces the amplitude of the signal from its original value. In case of PLC, when the high frequency signals injected into the power line by using an appropriately designed high-pass filter, maximum signal power is received only when the impedance of the transmitter, power line, and receiver match.

If we use a dedicated communication channels like Ethernet cable which acts as a channel to carry information in the network means it has known impedance so there is no attenuation problem in it. But, in the power line networks a variety of conductors were used and cross sections joined randomly. So in this case, a wide variety of characteristic impedances are encountered in the network.

Further, a network terminal impedance tends to vary both with communication signal frequencies and with time, depending upon the load pattern at the consumer's premises. This impedance mismatch results in a multipath effect characterized by deep notches associated with destructive interference at certain frequencies. In a typical home environment, the attenuation on a typical power line is 20 to 60 dB and depends heavily on the load.

3.4.2 Noise

The major sources of noise in the PLC channel are due to electrical appliances. It utilizes the frequency of 50 Hz or 60 Hz electric supply and generates noise components extending well into the high frequency spectrum as harmonics of the line frequency. Apart from these, induced radio frequency signals from broadcast, commercial, military, citizen band, and amateur stations severely impair certain frequency bands on the PLC channel. Electrical appliances can be divided into three categories based on the nature of the noise they produce in the high frequency bands. They are impulsive noise, periodic impulsive noise and continuous noise. Whenever electric switch is turned on or off it produce certain noise called impulsive noise.

When the source is used as triac controlled light dimmers, it introduces noise in between the AC path line for every AC cycle. This occurs repeatedly for every half of the AC signal. So it is called periodic impulsive noise. Continuous impulsive noise produced when the series wound AC motors are used. Such motors are found in electrical appliances like vacuum cleaners, drilling machines, electric shavers, and most of the common kitchen appliances. Commutator arching from these motors produces impulses at repetition rates in the range of several kilohertz. Continuous impulsive noise is the more severe than all other noise sources. High bandwidth digital devices communicating on the power line need to use powerful error correction coding along with appropriate modulation techniques to cope with these impairments.

3.4.3 Electromagnetic Compatibility

The use of power lines for communication is interesting to reduce the number of cables flying over. The information modulated with carrier frequencies at 9 kHz to 30 MHz range are transmitted in PLC. The skin depth effect at high frequency range causes the power lines to radiate high frequency electromagnetic signals. Skin effect is the tendency of an alternating electric current (AC) to become distributed within a conductor such that the current density is largest near the surface of the conductor, and decreases with greater depths in the conductor.

At high frequencies, the skin depth becomes much smaller. To avoid this effect the placement of any wireless service near PLC systems is bound to be subjected to interference. The interference is directly proportional to the transmission power and the distance between the installation and the power line. This calls for cautious design of filters to prevent the leakage of high frequency signals. The solution that integrates the 802.11b wireless networking protocol (Wi-Fi) with PLC is a typical case in which mitigating the interference plays on important role before deploying the equipment.

3.5 MODULATION TECHNIQUES FOR PLC

In PLC technology, how does the communication work? In communication technology, the sender modulates the data which is to be sent, then injects it into the medium, and the receiver demodulates the data to read it. Modulation is done to strengthen the signal to carry over long distances. When it comes to the PLC communication, we have to consider the need of modulation technique and then what type of modulation suits best for the communication should be examined before transmitting data.

The major difference is that PLC does not need extra cabling and it reuses the existing power wiring. Considering the pervasiveness of power lines, this means with PLC, and virtually all line powered devices can be controlled or monitored. Communication technology would be discussed by often referring to the 7-layer OSI model. Some PLC chips can implement only the physical layer of the OSI model, while others integrate all seven layers. One could use a digital signal processor (DSP) with a pure software realization of the MAC and an external PHY circuit, or an optimized system on chip (SoC) solution, which includes the complete PLC MAC and PHY. The Cypress CY8CPLCXX series is an example of the latter, with a ready-to-use physical and network layer, and a user-programmable application layer. Before moving on to the applications of PLC, let's first understand the various aspects of the physical layer by viewing it as three segments on the basis of data rates given in Table 3.1.

Table 3.1. PLC technology classification

Data rate	Low data rate 0-10 kbps	Medium data rate 10 kbps-1 Mbps	High data rate > 1 Mbps
Modulation	BPSK, FSK, SFSK, QAM	PSK + OFDM	PSK + OFDM
Standards	IEC 61334, ANSI / EIA 709.1, .2, UPB	PRIME, G3, P1901.2	G. hn, IEEE 1901
Frequency range	Up to 500 kHz frequency	Up to 500 kHz	In MHz
Applications	Control and command	Control and command, voice	Broadband over powerline, home networking

3.5.1 Modulation Schemes

A variety of modulation schemes can be applicable in PLC. Orthogonal frequency division multiplexing (OFDM), binary phase shift keying (BPSK), frequency shift keying (FSK), spread-FSK (S-FSK) and proprietary schemes too (for example, differential dode shift keying (DCSK) from Yitran) are a few modulation techniques used in the PLC. In Table 3.2, BPSK, FSK, SFSK and OFDM are compared on the basis of two important criteria, i.e., bandwidth efficiency and complexity (cost).

Table 3.2. Comparison of modulation schemes

Modulation scheme	Bandwidth efficiency	Complexity
BPSK	Medium	Low
FSK	Medium	Low
SFSK	Low	Medium
OFDM	High	High

OFDM is applied in high data rates, but it requires computational horsepower to churn out fast Fourier transforms (FFT) and Inverse FFT (IFFT), as required by the scheme. On the other hand, BPSK, FSK are robust and simple but offer lower data rates. The current trend is to move towards OFDM with PSK modulation (G3 and probably P1901.2). Such heavy computation requires DSP capability, whereas FSK, PSK and SFSK can be accomplished by a microcontroller.

Two distinctly different sets of standards apply to power line networking as of early 2010. Within homes, the HomePlug AV and IEEE 1901 standards specify how, globally, existing AC wires should be employed for data purposes. The IEEE 1901 includes HomePlug AV as a baseline technology, so any IEEE 1901 products are fully interoperable with HomePlug AV, HomePlug GreenPHY, and HomePlug AV2. On the other hand, medium-frequency home control devices remain divided, although X10 tends to be dominant. For power grid use, IEEE approved a low frequency (≤ 500 kHz) standard called IEEE 1901.2 in the year 2013.

Various standards listed in Table 3.3 have been developed in order to ensure reliable communications and interoperability, especially for the smart grid and home networking. Examples of such standards are:

- G3PLC (3rd generation PLC was developed to meet the industry's need for a potentially ubiquitous power line communications standard that will enable the smart grid vision. It facilitates high-speed, highly reliable, long-range communication over the existing power line grid.)

- PRIME (power line intelligent metering evolution) is a mature, consolidated and worldwide PLC standard for advanced metering, grid control and asset monitoring applications.
- P1901.2 Standard for low frequency (less than 500 kHz) narrowband power line communications for smart grid applications via alternating current and direct current electric power lines.
- The ANSI/EIA-709.1 standard, developed by an American company Echelon in the late 1980s under the name LonTalk, defines a network protocol which was originally suited to the demands of applications for building and home automation.
- KNX is an open standard for commercial and domestic building automation. KNX devices can manage lighting, blinds and shutters, HVAC, security systems, energy management, audio video, white goods, displays, remote control, etc.
- IEC 61334, known as distribution automation using distribution line carrier systems, is a standard for low-speed reliable power line communications by electricity meters, water meters and SCADA.

Table 3.3. Specifications of narrowband PLC standards

Standard	Technology	Frequency band	Bit rate (kbps)
G3-PLC	OFDM	36-90.6 kHz	5.6-45
PRIME	OFDM	42-89 kHz	21.4-128.6
IEEE P1901.2	OFDM	9-500 kHz	Coming soon
ANSI/EIA 709.1, .2	BPSK	86, 131 kHz	3.6-5.4
KNX	S-FSK	125-140 kHz	1.2
IEC61334	S-FSK	CENELEC-A	2.4

These, along with the organizations that govern them like CENELEC, FCC, ARIB, Homeplug Power Alliance specify ranges for operation of PLC. If a worldwide standard for PLC were to be established, this would have a positive impact on adoption of PLC. So far, the G3-PLC standard is touted as the most robust scheme available, and the IEEE 1901.2 working group is committed to developing a universally acceptable standard.

3.6 WIRELESS COMMUNICATION

Data communication can also be performed in wireless mode. This is a broad term that incorporates all the procedures and forms of connecting and communicating between two or more devices without wire via wireless communication technologies and its supporting devices.

Features of Wireless Communication

The evolution of wireless technology has brought us to stand in smart way by providing many effective features.

- The transmitted distance can be anywhere between a few meters (for example, a television's remote control) and thousands of kilometers (for example, radio communication).
- Wireless communication can be used for cellular telephony, wireless access to the Internet, wireless home networking, and so on by replacing the existing wired networking.
- Other examples of applications of radio wireless technology include GPS units, garage door openers, wireless computer mice, keyboards and headsets, headphones, radio receivers, satellite television, broadcast television and cordless telephones.

3.6.1 Wireless Networks

A wireless network enables people to communicate and access applications and information without wires. This provides freedom of movement and the ability to extend applications to different parts of a building, city, or nearly anywhere in the world. Nowadays, wireless networking is turned to the new face of networking. Wireless networking has been around for many years. Cell phones are one of the type of popular wireless communication for talking to each other worldwide.

Wireless networking is not only less expensive than more traditional wired networking but also much easier to install and use. A computer uses wireless data connections between network nodes. As with networks based on wire, or optical fiber, wireless networks convey information between computer devices. The information can take the form of e-mail messages, web pages, database records, streaming video or voice.

In most cases, wireless networks transfer data, such as e-mail messages and files, but advancements in the performance of wireless networks is enabling support for video and voice communications as well. Examples of wireless networks include cell phone networks, wireless local area networks (WLANs), wireless sensor networks, satellite communication networks, etc.

3.6.2 Types of Wireless Networks

3.6.2.1 Wireless Local Area Networks (WLANs)

WLANs given in Figure 3.7 allow users in a local area, such as a university campus or library, to form a network or gain access to the Internet. A temporary network can be formed by a small number of users without the need of an access point given that they do not need access to network resources. It supports D2D communication applications in IoT. In this network any number of users can be used and easy to add and remove from the network.

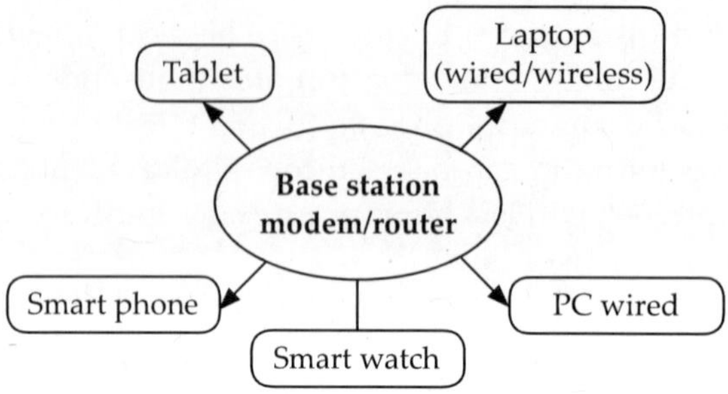

Fig. 3.7 WLANs

3.6.2.2 *Wireless Personal Area Networks (WPANs)*

The two current technologies assisting for wireless personal area networks are infrared (IR) and Bluetooth (IEEE 802.15) shown in Figure 3.8. These will allow the connectivity of personal devices within an area of about 30 feet. However, IR requires a direct line of site and the range is less.

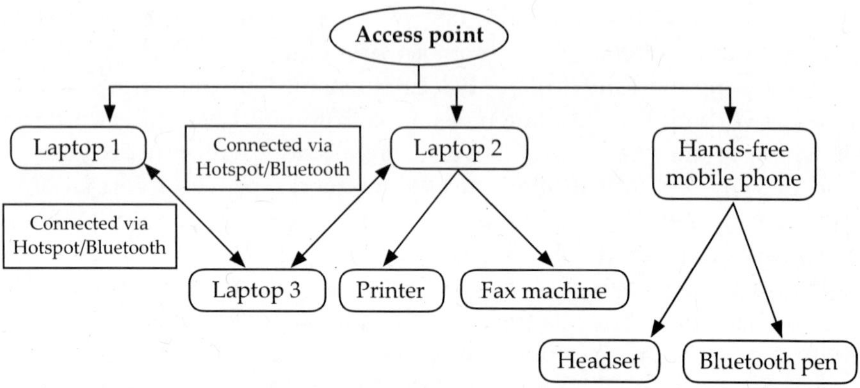

Fig. 3.8 WPAN

3.6.2.3 *Wireless Metropolitan Area Networks (WMANs)*

This technology allows the connection of multiple networks in a metropolitan area such as different buildings in a city, which can be an alternative or backup to laying copper or fiber cabling. Numbers 1, 2, 3, 4, 5 in Figure 3.9 are five different networks connected to the system.

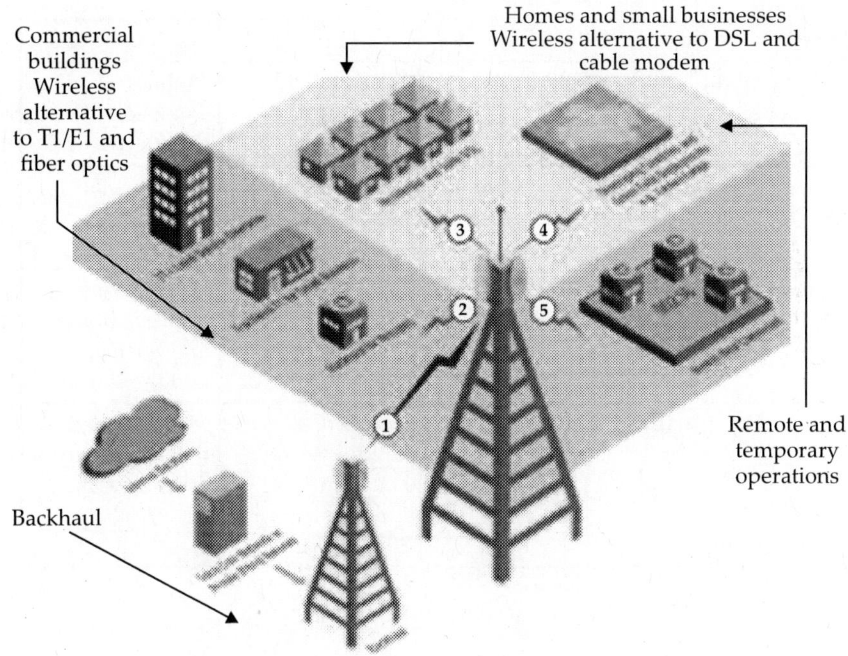

Fig. 3.9 WMAN
(https://cs.hac.ac.il/staff/martin/Wireless/slide05.pdf)

3.6.2.4 Wireless Wide Area Networks (WWANs)

These types of networks can be maintained over large areas, such as cities or countries, via multiple satellite systems or antenna sites looked after by an ISP. These types of systems are referred to as 2G (2nd generation) systems, a model of this system is given in Figure 3.10.

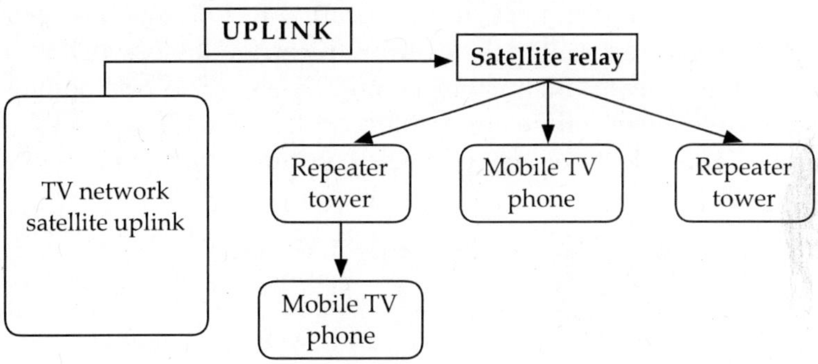

Fig. 3.10 WWANs

Table 3.4. Comparison of wireless networks

Type	Coverage	Performance	Standards	Applications
Wireless LAN	Within a building or campus	High	IEEE 802.11, Wi-Fi, and HiperLAN	Mobile extension of wired networks
Wireless PAN	Within reach of an individual workspace	Moderate	Bluetooth, IEEE 802.15, and IrDa	Cable replacement for peripherals
Wireless MAN	Within a city	High	IEEE 802.16, and WIMAX	Fixed wireless between homes and businesses and the Internet
Wireless WAN	Worldwide	Low	CDPD (Cellular Digital Packet Data) and Cellular 2G, 2.5G, 3G and 4G.	Mobile access to the Internet from outdoor areas

3.7 TECHNOLOGIES FOR WIRELESS NETWORK

The term wireless refers to the communication or transmission of information over a distance without requiring wires, cables or any other electrical conductors. The communication is set and the information is transmitted through the air, without requiring any cables, by using electromagnetic waves like radio frequencies, infrared, satellite, etc. At the end of the 19th century, the first wireless communication systems were introduced and the technology has significantly been developed over the intervening and subsequent years. Today, the term wireless refers to a variety of devices and technologies ranging from smart phones to laptops, tabs, computers, printers, Bluetooth, etc.

In recent days, the wireless communication technology has become an integral part of several types of communication devices as it allows users to communicate even from remote areas. The technology is used for wireless communication like cordless telephones, mobiles, GPS units, ZigBee technology, wireless computer parts, and satellite television, etc.

3.7.1 Bluetooth Technology (IEEE802.5.1)

Bluetooth technology, (Figure 3.11) allows the users to connect to a variety of different electronic devices wirelessly to a system for the transfer and sharing of data. Cell phones are connected to hands-free earpieces, wireless

Fig. 3.11 Bluetooth

(https://quantumcomputingtech.blogspot.com/2018/06/bluetooth-technology-images.html)

keyboard, mouse and mike to laptops with the help of Bluetooth, are in regular use. It transmits information from one device to other device with paired frequency.

Bluetooth works by the simple principle of sending and receiving data in the form of radio waves. Every Bluetooth enabled device has a card-like attachment known as the Bluetooth adapter. It sends and receives data. It has a particular range of connection. It works under the frequency of 2.45 GHz, the connection range is classified into three classes.

- Class 1 devices transmit at 100 mW with a range of 100 meters or 328 feet.
- Class 2 devices transmit at 2.5 mW with a range of 10 meters or 33 feet.
- Class 3 devices transmit at 1 mW with a range of less than 10 meter.

Most Bluetooth headsets and headphones are common Class 2 devices Bluetooth technology has many functions, and it is used most commonly in wireless communications market.

According to the official Bluetooth website, Bluetooth uses a low-power signal with a maximum range of 50 feet with sufficient speed to enable transmission of data. The pairing process identifies and connects any two devices to each other. It also prevents interference from other non-paired Bluetooth devices in the area.

3.7.2 ZigBee (IEEE 802.15. 4)

ZigBee (Figure 3.12) is a wireless communication standard module designed to address the unique needs of low power, low cost wireless sensor, and control networks. It can be used almost anywhere, as it is easy to implement and requires little power to operate. It has been developed

Fig. 3.12 Zigbee module

looking into the needs of the communication of data with a simple structure like the data from the sensors.

Its low power consumption limits transmission distances to 10-100 meters line-of-sight, depending on power output and environmental characteristics. It can transmit data over long distances by passing data through a mesh network of intermediate devices to reach more distant ones. It is mostly used in low data rate applications that require long battery life and secure networking. In case of large distance coverage extended ZigBee can be used which provides coverage up to 40 miles. The XBee XTender wireless bridge module is capable of extending the range of ZigBee.

It uses 128-bit symmetric encryption keys for secured data transmission. It has a data rate of 250 kbit/s with a frequency of 2.4 GHz ISM band, best suited for intermittent data transmissions from a sensor or input device. It is used in commercial applications like sensing and monitoring applications. It gives flexibility to do more with the reliable wireless performance and battery operations.

3.7.3 Wi-Fi (Wireless-Fidelity) (IEEE802.11n)

Wi-Fi (Figure 3.13) is the name of a popular wireless networking technology that uses radio waves to provide wireless high speed Internet and network connections. A common misconception is that the term Wi-Fi is short for "wireless fidelity," however this is not the case. Wi-Fi is simply a trademarked phrase that means IEEE 802.11x. A typical wireless router in an indoor point-to-multipoint arrangement using 802.11n and a stock antenna might have a range of 50 meters (160 ft) or less. Its frequency range is 2.4 GHz and 5 GHz.

Wi-Fi is a form of low-power wireless communication used by many electronic devices such as laptops, systems, smart phones, etc.

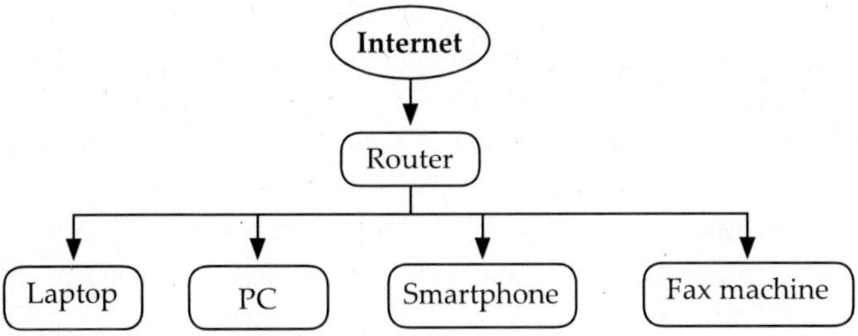

Fig. 3.13 Wi-Fi

In a Wi-Fi setup, a wireless router serves as the communication hub. These networks are extremely limited in range due to low power of transmissions allowing users to connect only within close proximity to a router or signal repeater. Wi-Fi is common in home networking applications which provides portability without any need of cables. Wi-Fi networks need to be secured with passwords for security purposes in order not to be accessed by others. The Wi-Fi network provides following advantages.

- **Ease of integration and convenience:** The wireless nature of such networks allows users to access network resources from nearly any convenient location.
- **Mobility:** With the emergence of public wireless networks, users can access the Internet even outside their normal working environment.
- **Expandability:** Wireless networks are capable of serving a suddenly increased number of clients with the existing equipment. In a wired network, additional clients require additional wiring.

3.7.4 WiMAX (IEEE802.16)

There are wireless broadband systems that offer fast web surfing without being getting connected through cable or DSL (example of wireless broadband is WiMAX). Although WiMAX can potentially deliver data rates of more than 30 Megabits per second, yet the providers offer average 6 Mbps and often deliver less, making the service significantly slower than the hard wired broadband. It provides coverage up to 50 km with a frequency range of 10-66 GHz (802.16a), and 2 to 11 GHz (802.16d). The actual cost of the data available using WiMAX widely varies with the distance from the transmitter. WiMAX is also one of the versions of 4G wireless available in phones as Sprint's 4G technology.

3.7.5 Satellite

Satellite communication is one of the broad wireless technologies, which is widely spread all over the world allowing users to stay connected virtually anywhere on the earth. The base stations used in this mode of communication, communicate directly with the orbiting satellites via radio signals. Portable satellite phones and modems have more powerful broadcasting abilities than the cellular devices as they have high range, apart from being more expensive in terms of cost, than their counterparts.

For example, for outfitting a ship through satellite communication, a traditional communication system is linked to a single satellite, which allows multiple users to share the same broadcast equipment. Wireless networking technologies connect multiple computers, systems and devices together without requiring wires or cables: a wireless local area network or WLAN comes under Wi-Fi.

3.8 SUMMARY

Various techniques used in IoT were discussed in this chapter. How the enhancements come from normal communication to smartness was given along with delivering detailed information about PLC in IoT. Then various wireless networks along with their corresponding available technologies which will help to automate any desired applications and its requirements were discussed.

REVIEW QUESTIONS

1. What is PLC?
2. What are the types of PLC?
3. Differentiate broadband over telephone lines.
4. Classify PLC over AC and DC lines.
5. Describe the importance of attenuation and electromagnetic capability in PLC.
6. Describe some modulation technique used in PLC.
7. What is meant by wireless network? Give some examples.
8. Explain the operating principle of wireless network in detail.
9. List the technologies used in wireless network.
10. Describe the short range wireless standards.

4

ARCHITECTURAL VIEW

4.1 INTRODUCTION

This chapter discusses the view of architectural structure of sensors, gateways, cloud and edge analytics. Its various standardization (national and international) levels like IEEE, IETF, ITU-T, OASIS, TSDSI and ETSI are briefed. It also explains the types of IoT testing and interoperability in device, network syntactical and semantic level.

4.2 IoT ARCHITECTURE

Construction of IoT architecture is not a unique or standard consensus. IoT architecture can be developed for the particular solutions. This development should be provided with flexibility (easy to add new functions or devices), integration with the existing system and interoperability. So the architecture will differ for different functions and solutions. But the functional area of IoT architecture consists of four basic stages as shown in Figure 4.1. They are physical devices (sensors/actuators), data acquisition and gateways, edge analytics and Cloud stages. The things or devices are equipped with either sensors for collecting data or actuators for performing commands. These are connected either wired or wireless mode. Gateways are used for preprocessing, data filtering and forwarding to the fog computing or edge analytics. Edge analytics is used for manual data processing and forward to Cloud. The Cloud stage is designed to store all kinds of information related to the network.

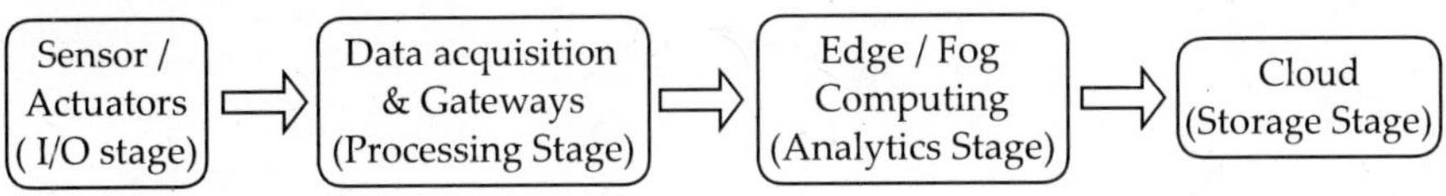

Fig. 4.1 IoT architecture

4.2.1 Sensors

Sensors play a role to get the physical parameter from the outside world or from the object itself as the input to drive the applications by the IoT system in smart way. Sensor is one of the important input devices in IoT structure to feed the data for data analysis. These sensors are either embedded in the smart devices themselves or implemented as a standalone objects to measure and to collect the data. Various sensors are available which are listed in chapter 2. For example, as shown in Figure 4.2, the different sensors connected to the system like the temperature sensor is to measure room temperature, soil sensor is to measure moisture level, fuel sensor is to measure the level of fuel in the vehicle, etc.

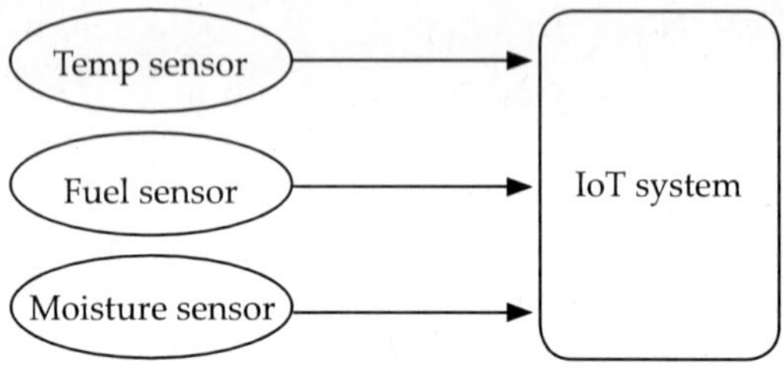

Fig. 4.2 Sensors interfaced with IoT system

Actuators

It is one of the indispensable elements which are used to transform the generated data by smart objects into smart action through the controller. For example, suppose the high temperature is sensed then it will pass the command to the actuators to turn ON the air conditioner to make the room cool as in Figure 4.3(a). The other example after sensing the data from the moisture sensor, if they need water then commands the actuators to open the selected values to pour the water to the soil as in Figure 4.3(b). The readings are continuously monitored till the defined value from the sensor is obtained. Then it will automatically goes to OFF by the feedback signal. All these processes are done without human intervention. That's why the system is called smart system or automated system.

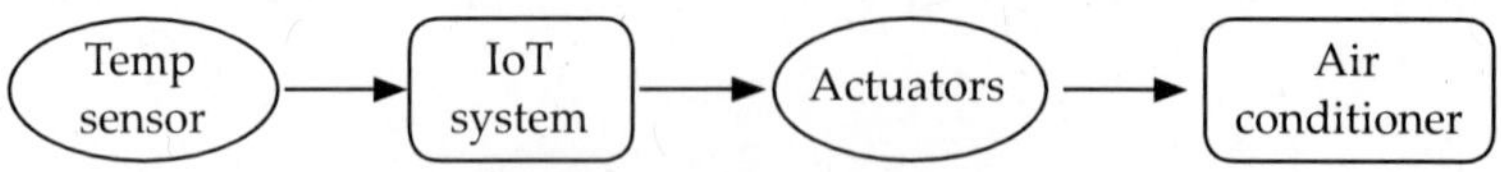

Fig. 4.3(a) Smart air conditioner

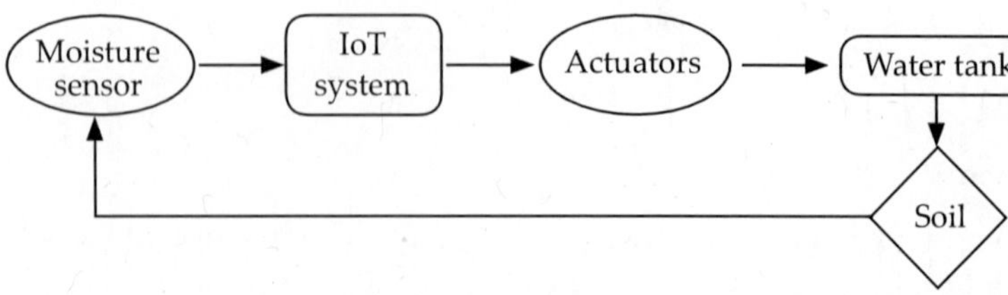

Fig. 4.3(b) Smart agriculture

Transferring the sensor data to the controller and then controller to actuators for action are important that should be done with precisely, secured without delay and are performed by the support of light-weight communication protocols. This protocol supports bidirectional communication with their corresponding gateways or data acquisition systems. It also supports to fit for the purpose, i.e., machine to machine communication with low power light-weight for any IoT devices needed with many users. Every objects should be able to recognize and can communicate for information sharing and also for collaboration in real-time applications. It is a resource constrained and battery operated device that are not easy one to communicate which requires more computing power, consumes more energy and bandwidth.

4.2.2 Gateways and Data Acquisition

This stage lies between first stage of the device (sensors / actuators) and the next stage of edge infrastructure and Cloud based platforms. The main function of this stage is a collection of data from massive volume of inputs and outputs. These are used for generating data, aggregation, selection and also transportation of data. It is also called intermediate stage between the connected things and edge infrastructure. The acquired data from the sensors is converted into the formats which are easily processed by the controller and easily transferred with compressed low bandwidth. It is responsible for communication between the sensors and the next stage. It controls the flow of data, filters the unwanted and selects the data to be forwarded in minimum size for ease transport. It is one of the preprocessing places of sensor data which makes bundles of data ready for further processing and providing data flow in both directions. It also provides end-to-end security supported with proper encryption and also prevents the leakage of Cloud data as well as reduces the risk of malicious attacks on the IoT devices.

Source devices cannot disconnect directly from the common platform and it requires IoT gateway to bridge the local environments and the existing platforms. Gateways are also to be designed in wireless technologies which cannot provide direct connection to the user network or the Cloud. The gateway reads the required information and sends the data to the common platform using a backhaul connection like cellular and Wi-Fi.

4.2.3 Edge Analytics

The information is generated or gathered at the device level and then communicated to the Cloud without the support of edge analytics. It creates more issues in the data analytics. So this edge analytics plays a vital role in the IoT architecture. It is placed before the Cloud, playing significant role on these factors like delay, regulatory issues, cost, scale and ease of operations.

This edge system provides quicker response and more flexibility in the processing and analysis of IoT data. The speed of the data analysis is important in industrial IoT to increase the productivity and quick decision making in the design. It is nothing but the hardware device and software gateways which are used to analyze and preprocess the data before transferring it to the Cloud.

4.2.4 Cloud

The Cloud stage is an important stage in the IoT architecture. It is also called data center. It is designed to store, process and analyze the massive volumes of data using machine learning mechanisms or data analytics engines. It reduces the storage on the system or device. Recently, the industrial IoT used the Cloud based system to increase the production rates, reduction of unplanned downtime and energy consumption. The Cloud can provide business intelligence and presentation options to help humans interact with system, control and monitor and also make informed decision on the basis of reports, dashboards and data observed in real time.

We can compare the IoT architecture with human body according to its main functions. The sensors are like neurons to get input; the gateway is the backbone of the IoT network to carry the information and the Cloud is the brain to store, to take decision and to provide control actions.

4.3 IoT STANDARDIZATION

Standardization is the key concept to accept the specifications and protocols universally. The IoT is the network of smart devices or things like sensors, actuators, network connectivity and software, used to sense data, process and store data. These data are exchanged between sensors and among the outside things. The things or smart devices should not only be addressable with one function but also be reconfigurable, reusable, locatable, uniquely identifiable and remotely accessible. The IoT is the growing field in the development of smart environment tremendously. The real-time IoT applications were implemented in remote health monitoring, crop monitoring, accident prediction and detection, traffic monitoring, pollution monitoring and unmanned aerial vehicles based reuse operation. Any kind of development should be approved by standard with the Govt. licensed organization or agency. It should be done before marketing the product or commercialization.

The emerging modern IoT technology is also used in educational entertainment. For that, we have to create standardization in architectural level, sensor level and networking level. Because of selected model it should be globally acceptable in the model wise and the components wise. Then only product will be commercialized in all countries. Global standardization of the IoT is the important solution in today's trend. Many national and international standard organizations are working together to bring the standard products within the country level and global level. Indian Standard Development Organization (SDO) is one of the country specific standardization for social uplifting and economic growth.

4.3.1 IoT Supports Organization Development

There is no single body for making IoT standardization in the current market. It is the responsibility of all national level and international level organizations at both private and public sector. So a general understanding has been formed by many domestic and multinational companies to accept common standardization and technology for the IoT. No global body is formed yet to standardize the IoT devices and connectivity like interoperability.

The leading organizations like IEEE, IETF, ITU-T, One M2M, 3GPP, etc. were actively involved in doing so at international level. TSDSI (Telecommunication Standardization Development Society, India), GISFI (Global ICT Standardization Forum for India), BIS (Bureau of Indian Standards), KATS (Korean Agency for Technology and Standards) and ETSI (European Telecommunication Standards Institute) were active at the regional level for IoT standardization.

4.3.2 International Level Standardization

Standard development organizations are ITU-T, IEEE, IETF, 3GPP and oneM2M have traditionally performed a pivotal role in defining technology standards to cover the overall problem space. They have specified either policies or reference architectures or have offered a standard protocol to carry out the communication. The role of each organization is discussed further in the following paragraphs.

4.3.2.1 IEEE Standardization

IEEE started their research for IoT standard in 2015 onwards. It follows that IoT refers to any system of interconnected physical objects and IT platform as well as any technology to better build, operate and manage the physical world via pervasive data collection, smart networking, predicative analytics, and deep optimization.

IEEE plays an important role in the forming of physical layer and medium access control (MAC) layer standards. It also keeps new specifications for producing IoT applications. The IEEE Standards Association (SA) undertook the project P2413 for IoT standardization. It mainly focused on architectural framework that covers needs for different applications and provides necessary technological solutions. IEEE P2413 considers the IoT as simple 3-tier architecture with applications, networking and data communication and sensing which are essential for the IoT communications. The IEEE standards for IoT are:

- IEEE 802.11: Wireless LAN is still a practical MAC standard for many applications.
- IEEE 802.15.4: LowPAN-low power sensing devices with low rate transmissions. It also made IP compatible through standardization efforts of IETF.

IEEE is also concentrating in defining several futuristic technology standards covering the lower layer specifications as well as application layer APIs in the specific domains of wireless access in vehicular environments (WAVEs), short range communication using visible lights and so on.

4.3.2.2 *IETF Standardization*

Internet Engineering Task Force (IETF) is a voluntary organization which developed the standard protocols for the Internet based network. They designed IoT-specific protocol for each layer. It also supports secured transmission in the transport layer by developing security protocols. These protocols are vary for the wired and wireless transmission. The wired transmission was supported by TLS (Transport Protocol Security) for Transport Control Protocol (TCP) and the wireless communication was supported by the support of DTLS (Datagram Transport Layer Security) for the User Datagram Protocol (UDP). The authentication process was done by the known key sharing or digital certificate sharing techniques that provided the secured transmission.

IETF associated with organizations like IEEE, 3GPP and ITU for developing and formatting the protocols of both physical layer and data link layer. Recently, IETF has created LPWAN standards for wide area applications in IoT network. The Internet protocol IPV6 is working over LPWAN group to optimize the IETF protocols for the different lower layer services to low power wide area network for SigFox, LoRA Alliance, 3GPP, etc. It is also concerned with IP compatibility for specific access technology. This representative network topology from LoRA Alliance is associated with LPWAN specific radio access technologies parallel to NB-IoT from 3GPP.

4.3.2.3 *ITU-T Standardization*

International Telecommunication Union (ITU) created a SG20 (Study Group) for enhancing IoT and its applications like smart cities and environments. Moreover, the study related to standardization extended towards big data analytics for bulk data, network requirement for IoT applications, allocating address and providing security. The ITU mainly offers the service towards IoT applications like health monitoring, smart agriculture, wearable devices and traffic monitoring and disaster preparedness. They focused on the component based reference model of IoT based M2M communication networked devices with or without the support of the gateways. It also defines the application level architecture for IoT.

4.3.2.4 M2M Standardization

In 2012, M2M was developed by global organization for creating global standards and global necessities of IoT based M2M technologies. They concentrated on specification of architecture, APIs, security, interoperability and certification for IoT devices used for M2M applications. Based on the applications it also published device layer architecture to interact with and exchange data between machines to machine and it offers a 3 layered network as given below.

- Application service layer
- Common service layer
- Network service layer

These service layers support and provide seamless message exchange and offer interoperability. They interact or exchange protocols bind with applications protocols like HTTP, CoAP and MQTT developed by M2M features.

4.3.2.5 3GPP Standardization

In 2016, 3GPP completed the standardization of NB-IoT which is a new narrowband radio technology developed for the Internet-of-Things (IoT). Recently it has announced 'LTE advanced Pro' specification and expecting essential backward-compatible changes. It took only 9 months to standardize the new technology after the study phase. Dino Flore, the Chairman of 3GPP RAN, said that once again 3GPP demonstrated the ability to quickly respond to the emerging market needs. With the completion of the NB-IoT, 3GPP has concluded a major effort in 'LTE advanced Pro' to address the IoT market. Portfolio of new technology 3GPP operators addresses their market requirements including NB-IoT, eMTC and EC-GSM-IoT.

- 3GPP Standard was set to be in 35% of low power wide area (LPWA) network connections by 2020.
- 3GPP supported a 'portfolio of technologies' that 3GPP operators can use to address market needs.

- **NB-IoT enhancements:** Agreement on NB-IoT positioning, support for multicast, power consumption and latency reduction, non-anchor PRB enhancements, mobility and service continuity enhancements, new power classes.
- **eMTC enhancements:** Support for positioning, support for multicast, mobility for inter-frequency, measurements, higher data rates and support for VoLTE.
- **EC-GSM-IoT enhancements:** Various radio interface enhancements.

3GPP has technical specification group (TSG) Chairs with SA/RAN/CT working group (WG) formed for the purpose to prepare, approve and maintain globally applicable technical specifications and technical reports for the following facilities:

- 3rd generation mobile system based on the evolved GSM core network, and the universal terrestrial radio access (UTRA)
- Further development of radio technologies such as LTE, coupled with evolution of core network elements such as the evolved packet core (EPC).
- Continuing evolution of the 2nd generation GSM/EDGE radio access network (GERAN), to be transposed by the organizational partners into appropriate deliverables (e.g., standards).

3GPP focused account of emerging ITU recommendations on interworking between IMT-2000 family members. In the framework of agreed relationships, the 3GPP technical specifications and technical reports will form the basis of standards or parts of standards of the organizational partners.

4.3.3 Standardization Efforts in National Level

The following organizations were standardizing the IoT domain in architecture level, network level, device level, protocol level and also its application level like TSDSI, GISFI, BIS, KATS and ETSI at the national level for IoT standardization.

4.3.3.1 OASIS Standardization

Organization for the Advancement of Structured Information Standard (OASIS) was a nationwide standard organization (SDO). MQTT and MQTT–S (sensors) were developed by IBM and operate over TCP/IP transmission purpose. They supported highly real-time low power MQTT-S mode for local exchanging operation on UDP. This protocol was used on the TCP layer for ensuring reliability and security. IBM brought the pair of protocol MQTT and MQTT-S under OASIS open standard community.

4.3.3.2 *TSDSI Standardization*

Telecommunication Standards Development Society, India (TSDSI) is an Indian SDO formed by the government of India to promote telecom standards in India. TSDSI is one of the eight organization partners of 3GPP and oneM2M for building cellular and IoT-related standards. TSDSI was Indian counterpart of other SDOs such as ETSI in Europe and ATIS in the USA. Currently, TSDSI has transposed the 3GPP and one M2M specifications as TSDSI technical specifications. TSDSI also represents India ITU-R and ITU-T for consolidating international efforts in the area of the IoT and telecommunications.

TSDSI has contributed low mobility large cell (LMLC) standard requirements to ITU-R. LMLC which is a unique requirement of developing geographies like India with large rural populations where a vast majority of people do not even have basic networking infrastructure. Unlike urban geography, rural areas have relatively low mobility, however, they are spread over large geographic area and hence require the larger cells to cover that entire region. The members of TSDSI have provided several other key contributions to 3GPP such as TDD based scheduling standard in future 5G networks.

4.3.3.3 *GISFI Standardization*

The Global ICT Standardization Forum for India (GISFI) is an Indian standardization body active in the area of information and communication technology (ICT) and related application areas, such as energy, telemedicine, wireless robotics, and biotechnology. The IoT reference architecture was explained in GISFI. It defines the following four layers as part of its generic architecture:

- *IoT device layer* includes individual sensors, network enabled objects, and capillary networks consisting of data sources that are nearer to the physical environment.
- *IoT gateway layer* consists of IoT gateways and connects to the IoT service platform layer through the core network; device and gateway layer functionality can coexist in a single device.
- *IoT service platform layer* defines different IoT service abstractions that can be used by multiple applications.

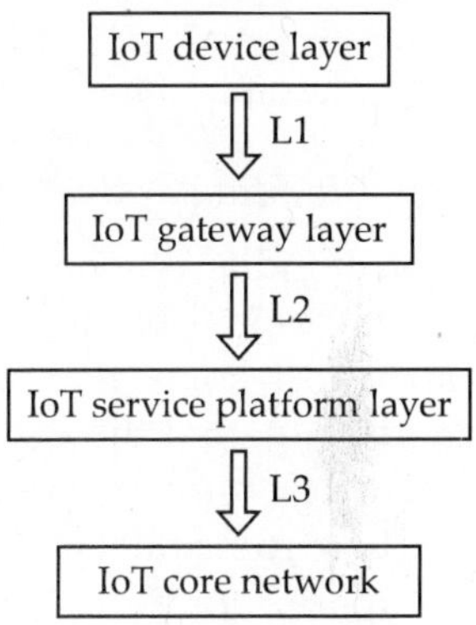

Fig. 4.4 IoT layers

- *IoT core network* is envisaged to be predominantly an IP based network. IP connectivity could be supported over multitudes of telecommunication infrastructures such as DSL, cellular networks (2G, 3G, 4G), and so on.

GISFI also identifies three reference points at the interfaces of these layers as shown in Figure 4.4 as follows:

(i) L1, interface from device layer to gateway layer
(ii) L2, interface from gateway layer to service platform layer through IoT core network
(iii) L3, interface from service platform to layer-specific vertical applications.

GISFI will be based on the concept of public-private partnership ventures. It is mainly focused on the following related application areas, such as energy, telemedicine, wireless robotics, and biotechnology.

- Unifying standardization efforts in India.
- Creating standards addressing the specificity of the Indian market.
- Creating an information center on Indian activities in ICT.
- Promoting Indian initiatives to develop into global standards.
- 'Vision To Be' is an ICT standards forum in India, which develops standards to meet the Indian requirements, as well as contributes towards the evolution of global standards.

India plays an important role in the processes of globalization in the field of the communication technology industry, research and development. GISFI addresses the research and product development of ICT in India and provides a bridge to the globalization of the Indian achievements, the issues of technology, governance, and development, and a platform for raising an awareness of the importance and the internationalization of the higher education in the field, supported by the joint partnership with the Government of India. The working groups organized by GISFI will draw information from academia, business, civil society, and government/policy making circles. It brings India's own procedure into line with what is normal elsewhere, so that cooperation is simpler and purchasing cheaper.

4.3.3.4 BIS Standardization

Bureau of Indian Standards (BIS) was the National Standard Body of India. BIS was responsible for the harmonious development of the activities of standardization, marking and quality certification of goods and for matters connected therewith or incidental thereto. The national economy is reached by providing safe, reliable and quality goods, minimizing health hazards to consumers, protecting the environment, promoting exports and imports substitute and controlling over proliferation of varieties, etc. The standards and certification scheme of BIS, apart from benefitting the consumers and industry, also supports various public policies especially in areas of product safety, consumer protection, food safety, environment protection, building and construction, etc.

In recent years, BIS has worked towards specifically addressing various national priorities and other government initiatives like Swacch Bharat Abhiyan, Digital India, Make in India and ease of doing business through its activities of standardization and certification. In standards development, BIS continues to address issues of technological changes and advancements, climate change, environment and energy conservation, conditions of health and safety and facilitation of trade. In the area of conformity assessment BIS is working towards making the processes simpler and faster. The main focus of this organization is harmonious development of the activities of standardization, marking and quality certification of goods to provide thrust to standardization and quality control for growth and development of industry on one hand and to meet the needs of consumers on the other.

The Bureau of Indian Standards Act, 2016, was implemented since 2017. The activities of BIS can be broadly grouped under standards formulation, international activities, product certification, hallmarking, laboratory services, training services, National Institute of Training for Standardization and Consumer Affairs and Publicity. The highlights of the new BIS Act are:

- Positions BIS as the National Standards Body.
- Allows multiple conformity assessment schemes in line with global practices.
- Enables the government to authorize any agency apart from BIS to certify and enforce conformity to a standard.
- Enables the government to include products under mandatory certification on grounds of health, safety, environment, national security and prevention of deceptive practices.
- Enables the government to bring hallmarking of precious metal articles under mandatory certification.
- Provides consumer protection measures like recall of non-conforming standard marked products, compensation to the consumer and more stringent penal provisions.
- BIS endeavors to create awareness and promote quality amongst all its consumers through various awareness programs.

4.3.3.5 KATS Standardization

In 1883, Korean Agency for Technology and Standards (KATS) was initially established as the Analysis and Testing. In 1999, as aligned under the Ministry of Commerce, Industry and Energy (MOCIE), KATS has set the position as representative national standardization body in Korea, overseeing various activities, development of Korean Industrial Standards (KS), control of quality and safety of consumer products, operation of legal metrology system, management of technical evaluation and certification of state-of-the-art technology and products, etc. In 2006, KATS has strengthened policy activities in standards and product safety and carried out a reorganization of division towards consumer-friendly and performance based administration system to actively engage in enhancing quality of life.

KATS was focusing on boosting national economic growth and being a platform of knowledge based society through development of strategic policies for evolving relationship between standards and technology, along with the relationship between standards and industry. KATS also seeks to create more convenient and safe environments via standardization.

It strives to adopt a new paradigm for standardization. It develops and implements a new bottom-up approach aimed at strengthening support for stakeholders from private sector to participate more actively in standards development process. KATS provides bureau of standard policy, bureau of product safety policy, bureau of conformity policy and bureau technical barriers to trade (TBT) affairs. The main objectives of KATS contributing to international standardization are:

- To provide technological infrastructure to support Korean enterprises to be recognized internationally by evaluation and certification of technology and quality.
- To enforce product safety.
- To improve quality of life through standardization.

4.3.3.6 ETSI Standardization

ETSI (European Telecommunication Standards Institute) was developed by the European Standards Organization (ESO) in 1988. It was a recognized regional standards body focused for telecommunications, broadcasting and other electronic communication networks and services. First they started to serve local (European) needs, but have a global perspective. Based on that it was one of the main standardization organizations for information and communication Technology (ICT) standards for European and global market needs.

- Being at the heart of digital
- Being global
- Being versatile
- Being inclusive

Each of these objectives is further detailed in our long-term strategy.

ETSI also provides members with an open, inclusive and collaborative environment for the international parties to work together. This environment supports them to think timely development, ratification and testing of globally applicable standards for ICT-enabled systems, applications and services. It is the forefront of emerging technologies across all sectors of industry and society that make use of ICT. ETSI shared its main objectives with their research group as being at the heart of digital, an enabler of standards, being global, versatile and inclusive. ETSI has listed some standards for IoT from 2012 onwards:

- **2012:** ETSI is a founding partner in the launch of the oneM2M partnership initiative for the global deployment of Machine-to-Machine (M2M) communications systems.
- **2012:** ETSI launches Cloud Standards Co-ordination at the request of the EC.
- **2015:** oneM2M release 1 specifications published.
- **2016:** oneM2M release 2 specifications published.
- **2017:** The Radio Equipment Directive (RED) completed its transition period supported by Harmonised Standards from ETSI.
- **2019:** First release of 5G ready for implementation.

4.4 IoT TESTING

Recently, the world development focused on IoT growing market with lots of opportunity. Today's IoT demand was urging to deliver better and faster services to access, create, use and share data from any device from anywhere. It needs trust to ensure the functionality and quality of the IoT devices. So testing is essential to ensure activeness. IoT testing is a kind of testing to check IoT devices and networks. IoT testing will ensure the quality of the device, web service and mobile. Along with that it has to support to validate the practical and non-functional requirements of the IoT solutions like security, interoperability and performance testing. This helps faster growth in time to market. IoT testing may be tough or challenging to certify in the devices, protocols, hardware, operating system, firmware, etc.

The main challenging issues of IoT were to find the adaptability for any communication with any device having different standards. It created a complexity of testing in quality of service, components and functionality. Advanced intelligence techniques have to be used for testing the hardware, software and connectivity. Everyday industries are focusing new challenges in IoT development based applications like collecting real-time responsiveness, availability, security, data privacy, big data analytics for data volume, variety, and veracity development of IoT protocols like MQTT, XMPP, CoAP handling of large number of sensors and its battery life.

4.4.1 Types of Testing

IoT testing is a type of testing to check IoT devices and its services. The thrust is to provide greater insight and control, over various interconnected IoT devices. Testing can be defined in architecture level or application level shown in Figure 4.5.

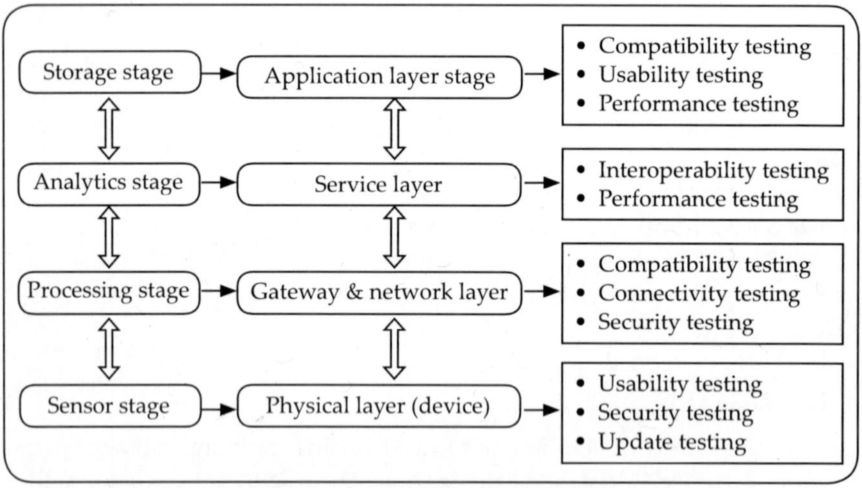

Fig. 4.5 IoT Testing framework

Generally, based on the functions, it has five types of testing, namely:

- Performance testing
- Security testing
- Usability testing
- Compatibility testing
- Interoperability testing

Let's describe each type one by one.

- **Performance testing:** Performance testing is to test the performance of all the things in the network. It had some approaches for developing and implementing communication network model along with its supporting embedded software system. This test has to be carried out at the application level and system level for data acquisition, processing, data analytics and network level for testing protocols like MQTT, XMPP, CoAP.

- **Security Testing:** IoT is a data-centric and connected system with its devices operated based on the database. The security should provide to data privacy, device identity, device authentication and decryption or encryption for data processing. Security testing is supported at the device level and protocol level to detect the source. This testing aims to secure the device, secure the networks and to secure the Cloud services by password protection and unique address to ensure the service.
- **Compatibility Testing:** IoT networks are connected with numerous devices with different configurations, operation of services with different supporting software. Compatibility testing is essential at network level and application level. It mainly validates the combination of different device version, supporting protocol version, communicating mobile devices and its supporting operating system (OS).
- **Usability testing:** This testing is used to test all functional blocks which are user friendly and customized in the development in the IoT applications. The IoT applications have many devices with many users with different perception. So it has to ensure the usability of each device like display data, processing data and confirmation of assigned job with its specifications.
- **Interoperability testing:** Interoperability testing is supported in session layer to enable a common platforms which support to communicate and programmable across devices regardless of make, model, manufacture or industry. It provided to test the ability to seamless interoperate protocols and device across different standards and specification. This testing makes sure that the device or network is suitable for any environment.

Real-time data testing

- **Pilot Testing:** It is software testing to a group of users was using the same software before the development of the software. This pilot testing mainly evaluates the feasibility, cost, time and the resources. It also finds the end user reaction towards the software, and its success rate. This system was exposed to a limited number of users in the real field before launching
- **Regulatory Testing:** Regulatory testing is one type of software testing to test the entire procedure step by step. It follows each testing with checklist for all regulations and compliances of each product with entire environment conditions. It supports to ensure the product performance by certificate testing with regulatory body for regulatory checklist.

  This testing supports to prepare for the safety and regulatory requirements in the segment of the product targets like home and building automation, transportation or healthcare. The wireless communication regulations were mainly helpful in the IoT applications where the products were communicated over the air.

- **Upgrade Testing:** Upgrade testing was one of the software testings to monitor the device with its updated software for compatibility and providing security in the device level and network level.

4.4.2 IoT Testing Tools

- **Virtual network simulator** is used to vary network load, connectivity, stability, etc.
- **Wireshark** is a GUI based open source used to monitor the traffic in the interface and host address.
- **TCP dump** is a command line based tool used to monitor the traffic and display the TCP/IP and other packets transmitted or received status over the network.
- **JTAG dongle** is a debugger. It helps to debug the target platform code.
- **Digital storage** oscilloscope (DSO) is used to check all events with time stamps, glitches in power supply and signal integrity check.
- **SDR** (software defined radio) is used to emulate receiver and transmitter for a large range of wireless gadgets.
- **Shodan** discovers the device information and its connected network and also used to track all device connections from the same Internet.
- **Thingful** is a search engine tool for IoT. It allows secure interoperability between more numbers of devices via the Internet. It also controls how data is used and empowers to take more decisive and valuable decisions.

4.5 INTEROPERABILITY

Nowadays more smart devices are found in the physical world which are interconnected and communicate with to each other via the existing Internet structure. It was supported to form global network infrastructure called IoT with combination of wired and wireless network with larger number of devices. So the interoperability issues arise due to cross platform, cross domain, and interfacing with non-interoperable IoT-devices while developing IoT-specific applications. Hence, it needs reliable and seamless communication between different vendors.

IoT, first coined by Kevin Ashton around 1999, had become an emerging technology in broad range of domains. The wide range of IoT applications were developed in various fields like healthcare, agriculture, commerce, transport and industrial, etc. These applications were developed by the leading vendors like Amazon, Cisco, IBM, Google and Qualcomm during the past few years.

These kinds of different vendors for different applications in a single smart device made to start the smooth operation by the term IoT interoperability. IoT interoperability is classified as five types as shown in Figure 4.6, namely, device, network, syntactical, semantic and platform interoperability to make bridge between two IoT devices. IoT interoperability research teams worked together for providing interoperability in services, application and platform level to IoT devices.

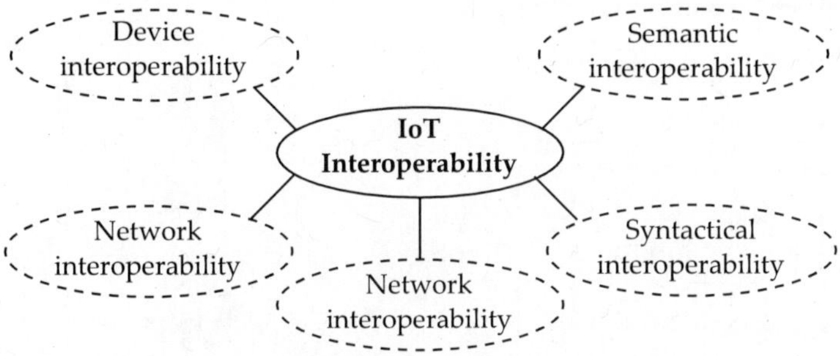

Fig. 4.6 IoT Interoperability

Interoperability research teams were formed for providing guidelines to operate in conjunction with other types to communicate according to all aspects. The main objective of the interoperability is to provide bridge between different specifications, data standards and middleware. The different types of interoperability are summarized in Table 4.1 and briefly explained as follows.

Table 4.1. Types of interoperability

Device interoperability (different devices)	Low-end device high-end device	Sensors smartphone, laptop
Network interoperability (different networks)	Heterogeneous network wired / wireless	Bluetooth, satellite, smartphone, MANET, WANET
Syntactical interoperability (different data format)	Data length, modulation, compression and encryption, parallel to serial, non-ASCII std to ASCII std	WSDL, REST APIs, XML
Semantic interoperability (different language)	Data model, information model	XML, CSV, JSoN
Platform interoperability (different OS)	Operating system, programming language, application field	Contiki, AWS IoT, linksmart, openIoT, Thingspeak HyperCat

4.5.1 Device Interoperability

IoT system consists of more number of smart things which can communicate with each other through the Internet. The connected smart things are either high-end devices or low-end devices as shown in Figure 4.7. The high-end devices are like active devices which have resources and computational capabilities. Example is smart phones and Raspberry Pi. The low-end devices are like passive devices which have non-computational and resource constrained in terms of energy and processing power. Example: RFID, sensors, actuators and Arduino.

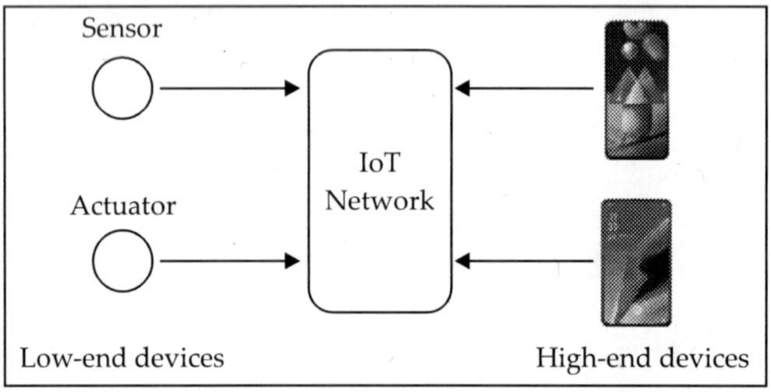

Fig. 4.7 Device interoperability

The device interoperability teams worked together to provide interoperation between heterogeneous devices. They should be focused on the characteristic of IoT devices like processor speed, RAM size, communication technology and battery capacity which is different of different brands and models. Nowadays, the smart devices like smart TV, wireless printer, smart air conditioner, etc., are operated with Wi-Fi or Bluetooth, Zigbee, Z-ware and also supported with non-standard solution like LoRA, SIGFOX.

Hence, different smart devices with different configurations have to work in the common platform for providing seamless service. Device interoperability is one of the important interoperability to enable the integration of more heterogeneous devices with various supporting protocols. It also supports the exchange of information between heterogeneous devices and between heterogeneous communications protocol which are able to integrate new devices with the existing IoT device.

4.5.2 Network Interoperability

IoT devices are communicated via wired and wireless network which have heterogeneous, multiservice, multivendor and large number of distributed network as shown in Figure 4.8. Different smart device signals can travel in anyone or more than a number of networks. So we can say that it has heterogeneous network which are with different range of operating frequencies either in long ranges or in short range communication devices. Hence, network interoperability term is important to make smooth or reliable connection between two smart devices for end-to-end communication for supporting interoperable between different networks. The compatibility should ensure in each node or router operating over different frequency ranges, different networks such as wired or wireless different network standard like 3G, 4G towards technology updating based on different service providers and different multiple access schemes. This might be developed via a supporting software or supporting hardware like adaptor, etc.

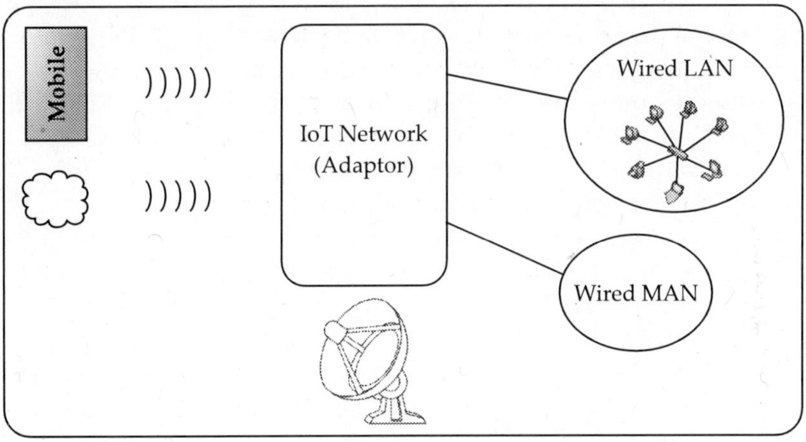

Fig. 4.8 Network interoperability

Conventional Universal Plug and Play (UPnP) and DLNA protocols are used for communication between IoT devices and the gateway. Few researchers proposes an IoT gateway based on user-space programmable software to bridge the heterogeneity between WSN protocols and mobile communication networks or Internet and includes functionalities like data forwarding, protocol conversion and management.

IoT devices use dissimilar communication techniques like Bluetooth, Zigbee, Satellite, etc. The IoT connection environment was dynamic, heterogeneous and over the air (OTA) type. So the network interoperable research teams should handle the issues like addressing, routing, resource optimization, security, OS and multi-support based on new era. The network operability was working in four different approaches to provide the interoperation. They are:

(a) IP based approaches
(b) SDN approaches
(c) NFV approaches
(d) Fog computing

(a) IP based approaches: While communicating through the Internet, IP plays a main role to connect two smart devices like sensors, gateways, etc. By implementing the TCP/IP stack on sensor nodes, gateways and protocol translations are not required. Due to the sensor nodes because of their resource constraint, the IETF has formed working groups (WGs) at the network layer such as Routing Over Low Power and Lossy Networks (ROLL), IPV6 over Low Power WPAN (6LoWPAN), Constrained Application Protocol (CoAP) which is based on UDP, and Constrained Restful Environment to solve the connectivity problem of resource-constrained devices.

(b) SDN approaches: Software Defined Networking (SDN) is a new networking paradigm to make the current wireless and mobile networks more intelligent, efficient, secure, and scalable in order to handle the large amount of data produced in the IoT. One of the main novelties of SDN for breaking the vertical slots in IoT, is to separate the control and data planes in networking devices.

SDN has been applied to IoT to facilitate networking applications such as heterogeneity, mobility management, QoS management, and security. SDN used to allow different objects from different networks to communicate with each other using IPV6 and at the same time simplify the management and control operations of various object types by adding an additional IoT controller over the SDN controller (Figure 4.9).

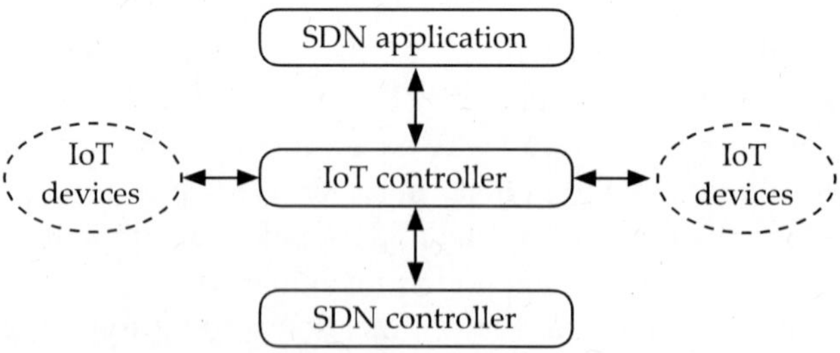

Fig. 4.9 IoT with SDN

Acquired signal from sensor was stored in the remote system through the gateway with Internet. During this transfer application, transport and network layers are involved with its protocols HTTP, TCP and IPV6, respectively. These can be updated between the gateways as COAP, UDP and 6 LOWPAN for the specific IoT devices as in Figure 4.10

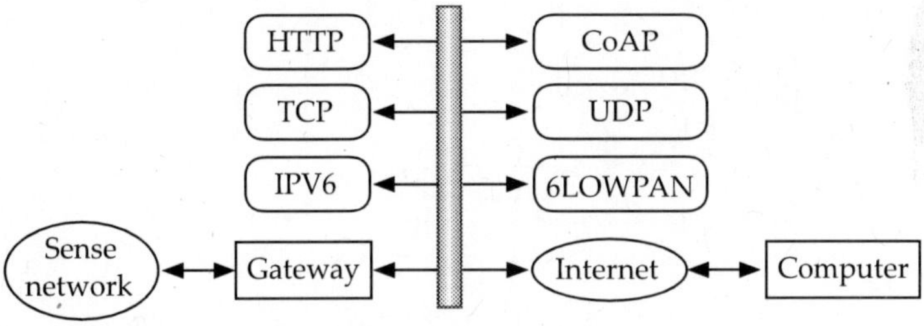

Fig. 4.10 Protocols in gateways

(c) **NFV approaches:** Network function virtualization (NFV) separates the physical network equipment (i.e., network address translator, Firewall) from the functions that run on them. NFV has the potential to reduce operational expenditure (OPEX) and capital expenditure (CAPEX) costs by sharing the network infrastructure, dynamic scaling, on-the-fly, and flexible network function deployment.

(d) **Fog computing:** The Cloud has been used as a medium to address interoperability called the Fog of Things, where the computing, storage and networking services are placed at the edge of the network rather than centralized Cloud servers, i.e., as close as possible to the end user devices. Fog computing paradigm provides value to the data before making it available to the web facilitating interoperability in IoT, 5G, AI, tactile Internet, virtual reality, and other complex data and network-intensive applications and preparing the managed data for further applications to be interoperable. Fog computing provides interoperability of local ecosystems in the fog and also at the Cloud level.

4.5.3 Syntactical Interoperability

Syntactical interoperability is mainly used to provide interoperability in the data format between two users. It focuses on smooth exchange of data handling in each device with different format, length, type, audio, video, etc. in data structure in the heterogeneous IoT network system as shown in Figure 4.11. Hence, some resources were developed to provide interoperation between the smart devices, for example, WSDL (web service definition language) and REST APIs suitable for interoperable. REST APT that allows the programmer to automatically convert and exchange data between HTTP, CoAP and MQTT. The message in parallel content is converted into serial, then exchanged via channel by XMT or JSON (JavaScript Object Notation) tool, compression/decompression tools and also includes encryption and decryption along with data to enhance the security. Some standard syntactical rules are used for encoding/compression at the transmitter side and the inverse of the same rules used for decoding/decompression at the receiver side.

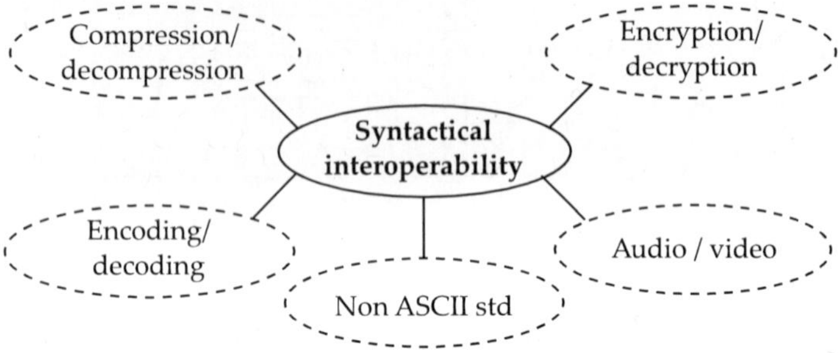

Fig. 4.11 Syntactical interoperability.

When the syntactical rules differ between the selected receiver and transmitter users for encoding and decoding, incompatibility occurs in the communication. This is called syntactical interoperability. By avoiding the mismatching of syntactic rules between the selected pairs for transmission can reduce the syntactical interoperability. Both users should follow the same syntactical rule for smooth transmission.

4.5.4 Semantic Interoperability

Semantic means relative to meaning in language or logic. Many smart devices do not speak the same language and cannot exchange various symbols across different gateways and smart hubs. The data may be represented in diverse units of measurements and other information. When two different data models or information models were exchanged dynamically or automatically, then semantic interoperability occurs. To minimize the semantic error, users have to avoid dissimilar data or incompatible format by different sources. It was supporting to enable interoperation between different agents, between different services and activating applications having different data.

Figure 4.12 shows that compatibility occurs because of different language. But this problem is made not by the machine which was working properly. This issue was caused by the human understanding skill level due to the lack of knowing new languages.

The semantic gateway as a service (SGS) is presented as a gateway between the physical world and the high-level layers of an IoT system. This SGS structure having more facility towards the sensor like the raw sensor data are transferred from external sink nodes to the central gateway node via the multi-protocol proxy. Before being forwarded, data are

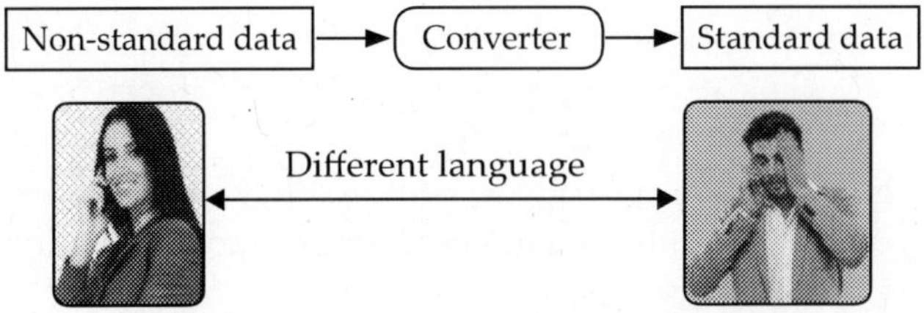

Fig. 4.12 Communicating with different languages

semantically annotated using W3C SSN ontology, SemSOS tool and other domain specific ontologies. Semantic annotation of sensor data provides semantic interoperability between messages and supply higher-level actionable knowledge for implementing.

The semantic web of things (SWoT) paradigm is proposed for the integration of the semantic web with the WoT, for realizing a common understanding of the various entities which form the IoT. Recent research has concluded that semantic web technologies are a major driver for interoperability across heterogeneous environments. Ontologies (or vocabularies) in IoT are a set of objects and relationships used to define and represent an area of concern. Many ontologies have been proposed in the context of IoT such as W3C semantic sensor network (SSN), IoT-ontology, SAREF and OpenIoT. The IoTivity focuses on device interoperability irrespective of form factor, operating system or service provider through protocol plug-ins.

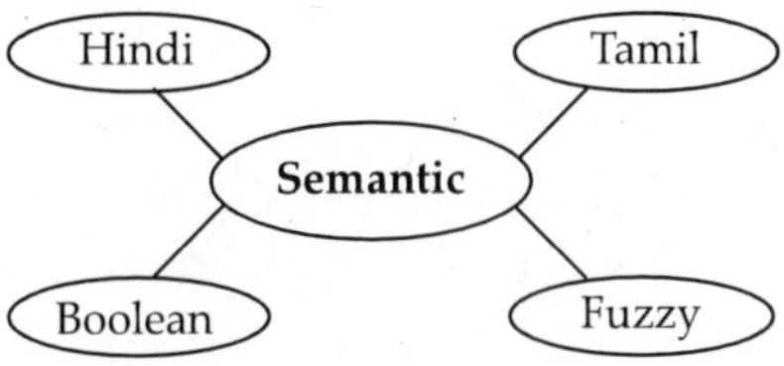

Fig. 4.13 Semantic interoperability

4.5.5 Platform Interoperability

Common platform means which will be suitable for all kinds like device, software and hardware. Here platform interoperability is defined as the availability of different operating system, different program language, different architecture and different accessing technique over the IoT things as shown in Figure 4.14. Recently more operating system software were developed for IoT devices like Contiki, RIOT, Tiny OS, open WSN, etc. These operating systems (OS) were differing from different services and different version. Sometimes the older version would not be compatible with new version along with its updates like difficulty in using Windows 2010 and Windows 2008. The Apple company products have unique device specification with its own operating system and its accessories. That kind of smart system will not be compatible with any other system like Samsung, Sony. But other company products are compatible within either software or its accessories.

Nowadays, most of the companies create their own OS and operate with their corresponding kits like Apple Homekit uses the open source language Swift, Google Brillo supports with Weave, Intel Galileo kits, Amazon AWS IoT system. Some of the system supported with embedded C, Rasberry Pi, Android and Python Language and Node JS / Node MCU / ESP32. This kind of non-uniformity between the product developers leads to think about to design for cross platform or cross domain by the IoT application developers. The IoT common platform developers should have knowledge about the complete history of the two different smart users/devices.

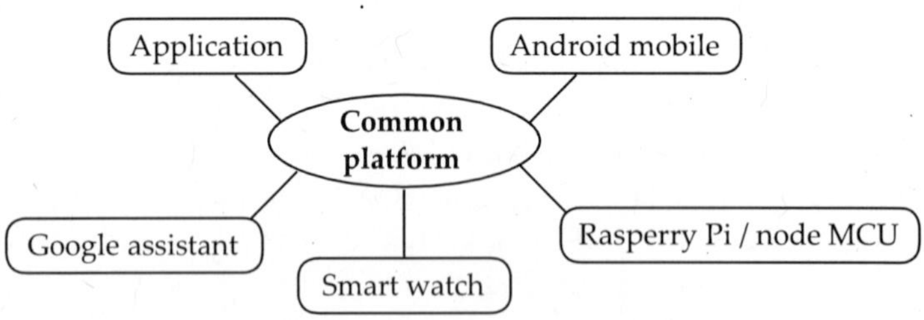

Fig. 4.14 Cross platform interoperability

The conventional sensors are not suitable for the medical applications. So it is needed to design a separate biosensor for sensing the bio-signals from the human body. The human body temperature or signals are varied from the electrical signal in its amplitude and phase. This platform

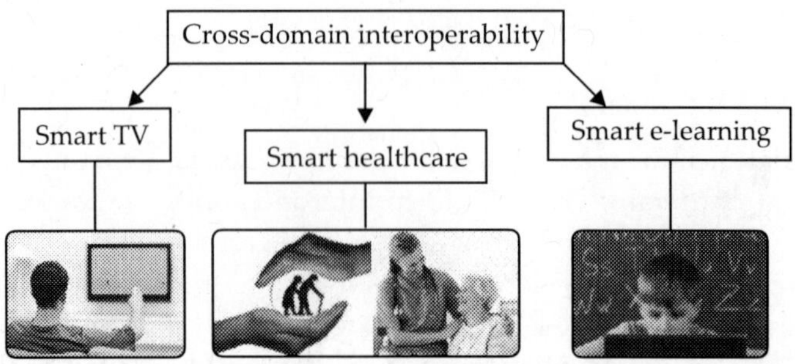

Fig. 4.15 Cross-domain interoperability

interoperability research teams should focus on enabling different devices with different services to make compatibility within the devices which are connected in different domain, or services vice versa different services should be compatible with different devices for same application. Figure 4.15 indicates the cross-domain operation in the IoT platform in home applications like watching TV, e-Learning for kids and smart healthcare enabled for the elder.

4.6 SUMMARY

From the previous chapters, introduction about IoT, its layer-wise devices, technical view was given. But in this chapter the overall architectural view along with important standardization and testing was discussed. In the forthcoming chapter communication and security part was discussed.

REVIEW QUESTIONS

1. To design IoT model, what are the domains needed to construct?
2. Write down the importance of IoT standardization.
3. Compare various standardizations of IoT.
4. What is the need of IoT testing?
5. What are the types of application level testing?
6. List some testing tools available for IoT.
7. Why have to adapt interoperability in IoT?
8. Explain semantic and platform interoperability.
9. Describe syntactical interoperability.
10. What are the approaches used in the network interoperability?

5

COMMUNICATION AND NETWORKING IN IoT

5.1 INTRODUCTION

IoT is an emerging technology to make our normal life a smart life. This book is an easy hands-on guide. Here, in this chapter layer-wise communication protocol, that is, set of rules that need to follow by each layer is given along with its features.

5.2 LAYERS OF COMMUNICATION PROTOCOLS AND ROUTING

The OSI layer of communication network consists of standard seven-layer structure. These seven layers are:

1. Physical layer.
2. Data link layer.
3. Network layer.
4. Transport layer.
5. Session layer.
6. Presentation layer.
7. Application layer.

These are the backbone of the network architecture. It is more correlated with IoT to develop the layers and its protocols. The layer-wise protocols are explained briefly as follows:

5.2.1 Physical Layer

Physical layer is an important layer for making communication between two users either in wired or wireless mode. The main purpose of this layer is to provide the physical device connections and collection of data in bit streams for further processing. It is directly connected to the data link layer. The IoT physical layer consists of an electronic embedded systems comprising of sensors, actuators and controllers.

5.2.2 Data Link Layer (DLL)

The data link layer provides a link between the physical layer and network layer by using a certain set of protocols. In IoT, depending upon the network access, various kinds of protocols are used. In this layer data transmission takes place in the form of frames. It depends upon the various Internet access having varieties of DLL protocols. They are:

(i) **IEEE 802.15.4e: Low Rate WPAN:** It uses time synchronization, channel hopping to enhance reliability. It is suitable for all IoT low power applications.

(ii) **IEEE802.11:** The Wi-Fi protocol has synchronization frames, efficient bidirectional packet exchange, short MAC frame, null data packet and increased sleep time. The only disadvantage in Wi-Fi (802.11) is that it does not support IoT due to frame overhead and high power consumption. It is used in digital devices, laptops, mobiles, etc. In order to overcome this IEEE802.11ah is evolved.

IEEE802.11ah - WLAN: It defines a WLAN system operating at sub-1 GHz license-exempt bands. Due to the favorable propagation characteristics of the low frequency spectra, 802.11ah can provide improved transmission range compared with the conventional 802.11 WLANs operating in the 2.4 GHz and 5 GHz bands. 802.11ah can be used for various purposes including large-scale sensor networks, extended range hotspot, and outdoor Wi-Fi for cellular traffic offloading, whereas the available bandwidth is relatively narrow. The protocol intends consumption to be competitive with low power Bluetooth, at a much wider range.

(iii) **Wireless Hart:** It uses TDMA (time division multiple access) in its MAC (media access control). It has separate encryption algorithm and provides high security and integrity. It uses in both wired and wireless IoT communication with high level security.

(iv) **Z Wave:** It provides point-to-point communication for small messages.

- Frequency: 908.42 GHz
- Range: 30-100 m
- Data transfer rate: 100 kbps

CSMA (carrier sense multiple access) CA (collision avoidance) mechanism for media access is used to provide high reliability, low power consumption, remote or local control, simple installation, interoperability, home automation which includes smart homes, small commercial domains, smart product based applications, energy saving and home security.

(v) **Bluetooth low energy:** It provides low latency and fast transmission,

- Standard: Bluetooth
- Frequency: 2.4 GHz
- Range: 50-150 m
- Data transfer rates: 3 Mbps power saving (nodes awake only communicating otherwise it sleeps)

It is wireless, economical, easy to install and free to use if the device is installed with it. The main disadvantage is that it supports only a short range communication network. It can connect only two devices at a time and is used mainly in vehicle networking applications.

(vi) **Zigbee smart energy:** It supports medium range wireless communication
- Standard: ZigBee based on IEEE802.15.4
- Frequency: 2.4 GHz
- Range: 10-100 m
- Data transfer rate: 250 kbps

It supports full mesh networking, security, scalability and provides better performance wireless mesh networking, direct communication, low power consumption but it is too costly and works with low speed within a small distance. The main applications are smart homes, remote controls, healthcare system, commercial and residential control, home networking, industrial control and management and consumer electronics.

(vii) **Das 7:** It supports encryption and IPV6 addressing. It is highly scalable, long range communication and provides outdoor coverage with high data rate and low cost. It is used in the active RFID devices.

(viii) **Home plug:** Home plug AV1 and Home plug AV2 covers both physical and MAC layers. It provides less power consumption home automation applications.

(ix) **G.9959:** It is a standard for short range narrowband digital radio communication transceivers. It supports low bandwidth, low cost, half duplex communication, high reliability and it has unique network identifiers, collision avoidance mechanism, automatic retransmission, to save power and delay sensitive application.

(x) **LTE-A:** It is scalable, cost efficient and uses OFDMA. It supports extended coverage, indoor positioning, broadcast and multicast in a single cell. It is mostly used in cellular networking applications.

(xi) **LORAWAN:** It is location free, low cost, low power, energy harvesting, security, bidirectional and suits for long range wide area network properties.
- Standard: LoRaWAN
- Frequency: 433 MHz – 923 MHz
- Range: 2-5 km (urban environment), 15 km (suburban environment)
- Data rates: 0.3-50 kbps.

(xii) **Weightless:** It has a two band. One is narrowband denoted as Weightless-N. Another one is wide band denoted as Weightless-W

Weightless-N: Uses ultra narrowband in sub 1 GHz ISM band.

Weightless-W: Uses television band frequencies and frequency hopping to minimize interference. It suits for wireless IoT applications.

(xii) **IEEE802.3 (Ethernet):** It supports the data rate of 10 Mbps to 40 Gbps. It uses coaxial and fiber optic cable for faster communication. It is used for wired network applications.

(xiii) **IEEE802.16 (Wimax):** It supports the data rate of 1.5 Mbps to 1 Gbps. It is used for wireless broadband connection.

5.2.3 Network Layer

The network layer is an essential layer of the OSI model. It acts as the backbone of the computer network. The main purpose of this layer is to provide the most effective logical path for information transfer between DLL and transport layer. This layer consists of hardware devices like routers, bridges, and firewalls. It creates a logical image of the foremost economical communication path and provides the communication way to the physical devices.

(i) **RPL (routing protocol for low power and lossy network):** As a distance based vector routing protocol, it uses DODCG (distance oriented directed cyclic graph) method and keeps tracking of nodes based on distance. It is used in IoT application based on routing.

(ii) **CORPL (Cognitive RPL):** Have multiple forwarder, best neighbor hop will be chosen to forward the packet. It is used in IoT application based on routing.

(iii) **CARP (channel aware routing protocol):** Routing is based on the distributed networking. It is a light-weight packet forwarding protocol. It does not support reusability of previously collected data, e.g., underwater communication.

(iv) **Enhanced CARP:** Designed to overcome the disadvantage of CARP, it saves the previously collected data.

(v) **6LOWPAN:** IPV6 over low power wireless PAN, suits for different length address, It supports different network topologies, low bandwidth, low power consumption. It encapsulate IPV6 long header in 802.15.4 small MAC frames. It provides mobility, reliability, long sleep time, low power wireless applications.

(vi) **6TiSCH [IPV6 over TSCH (time slotted channel hopping)]:** Here, long IPV6 is sent through 802.15.4e data link. It stores available frequency and time in matrix form. The main disadvantage is that scheduling becomes an optimization problem. It is used for the highly secured IoT applications.

(vii) **6LO (IPV6 over low power nodes):** It links IPV6 over network of resource constrained nodes and other protocols and used for applications with dynamic ranges.

(viii) **IPV4 (internet protocol version 4):** It identifies the devices using hierarchical scheme. It uses 32-bit addressing schemes. The main disadvantage is it that does not provide delivery guarantee. It is used for wireless long range IoT applications.

(ix) IPV6 (internet protocol version 6): It uses 128-bit address and employed in IoT wireless long range applications.

5.2.4 Transport Layer

The transport layer provides information transmission services between the network and session layer. This layer gets the information and processes it using some protocols. The various services are provided by the transport layers for performing point-to-point communication, maintaining reliability by checking frequent failure, getting multiple input and single output facility. It consists of two standard protocols like TCP and UDP.

(i) TCP(RFC793): The transport control protocol (TCP) is a connection oriented and stateful protocol. It maintains reliability, error detection, flow control, congestion control and web browser based applications.

(ii) UDP(RFC768): The user datagram protocol (UDP) is a connection-less protocol. The drawback of this protocol is that it provides guaranteed delivery, proper ordering of messages and duplicate eliminations. It is used in time sensitive applications.

Protocols of the data link layer, network layer and transport layers are summarized and listed in the Table 5.1

Table 5.1. Protocols in data link, network and transport layer

Data link layer	Network layer	Transport layer
– IEEE802.15.4e – IEEE802.11 – IEEE802.11 ah – Wireless Hart – ZWave	– RPL (Routing protocol for low power and lossy network) – CORPL (Cognitive RPL) – CARP (Channel aware routing protocol)	– TCP
– Bluetooth low energy – Dash 7 – Zigbee smart energy – Home plug – G.9959	– ECARP – 6LOWPAN – 6TISCH	– UDP
– LTE-A – LORAWAN – IEEE802.16 (Wimax) – IEEE802.3 (Ethernet) – Weightless	– 6LO – IPV4 – IPV6	

5.2.5 Session Layer

Session layer controls and manages the timing gap between transmission and reception of packets between the layers. In a network, the communication is not single way. It is the multi-directional communication and so in order to maintain all those communications, the special layer and management part is necessary that could be provided by the session layer.

(i) **MQTT (message queue telemetry transport):** It uses publish or subscribe architecture and it is a TCP based protocol. It provides link between the application and the user at one end and communication and network at other end. It suits for publish or subscribe type applications.

(ii) **SMQTT (secured MQTT):** It is the MQTT with added security. Here one message is encrypted and delivered to multiple other nodes.

(iii) **AMQP (advanced message queuing protocol):** TCP based protocol, uses both publish/subscribe, request/response methods. Used in financial industries runs over TCP.

(iv) **COAP (constrained application protocol):** It has low overhead REST HTTP interface. It is built over UDP and provides reliability, and duplication of message. It is applicable for web based applications.

(v) **XMPP:** It uses XML languages. It supports publish/subscribe, request/response architecture. It provides security, low latency for small messages. The main drawback of this protocol is that it could not provide the guaranteed quality of service. It increases power consumption due to additional overhead by XML. It is used for real-time applications.

(vi) **DDS (data distribution services):** It supports the publish/subscribe architecture. It has 23 quality of service level. It contains two important layers which are: data centric publish-subscribe, and data local reconstruction sub layer. It gives high QoS, high reliability and offers security, urgency, priority and M2M communications.

5.2.6 Presentation Layer

This layer presents the data to the application layer. Here, the data is checked for prescribed format. The encryption and decryption of data takes place here and data is translated as per the application needs for the user friendly access.

5.2.7 Application Layer

The user interfacing layer of the communication network is the application layer. It is usually connected to the global network for public access and private network for industry and corporate needs. The routing and management of the information can be done effectively. It consists of two protocols, namely, HTTP and Websocket.

(i) **HTTP:** It supports request/response model. It is a state-less protocol. It uses URI to identify resources. It is deployed for web applications having HTTP client as a browser.

(ii) **Web socket:** It allows the full duplex communication over a single socket and used in mobile application.

Table 5.2. List of next four layers protocols of IoT

Session layer	Application layer	Management layer	Security layer
– MQTT (Message queue telemetry transport) – SMQTT (Secured MQTT)	– HTTP (Hyper text transfer protocol)	– LWM2M – IEEE1905.1 (Interconnection of heterogeneous data link)	– Security within IoT protocol layers. – TLS/DTLS – Ubiquitous green computing control network security.
– AMQP (Advanced message queuing protocol) – XMPP	– Web socket	– TrR069 – OMA-DM (Open Mobile Allowance)	– TCG (Trusted computing group) – OAuth2.0
– DDS (Data distribution services) – COAP (Constrained application protocol)		– Smart transducer interface	– SASL (Simple authentication security lags) – ACE

Table 5.2 listed the protocols in session, application, management and security layer. In addition to the protocols in OSI seven layers, management layer and security layer protocols are included in IoT. They are briefly explained as follows.

5.2.8 Management of Protocols

Management protocol controls and does the flow maintenance in various IoT applications using following protocols.

(i) **IEEE1905.1 (interconnection of heterogeneous data link):** It allows different protocols to communicate with device without changing the system design. It establishes neighbor discovery, topology exchange, change notification, traffic measurement and security association.

(ii) **Smart transducer interface:** It offers management of different analog transducers and sensors. It supports the plug-and-play environment by TEDS (Transducer Electronic DataSheet). It's memory usage can be minimized by small XML messages.

(iii) **TR-069 (Technical Report 069):** It is a technical specification of the broadband forum that defines an application layer protocol for remote management of the customer premises equipment (CPE) to an IP network. Management is done by HTTP messages, employed in remote management of M2M devices.

(iv) **OMA-DM (Open Mobile Allowance):** XML message for communication is built over HTTP. Protocols are complex for resource constrained IoT devices. It provides remote provisioning, updating and managing faculty issue of M2M.

(v) **LWM2M:** It supports client server protocol. JASON (Java-script object notation) are used for communication. It is built on COAP. It is used in management of light weight M2M.

5.2.9 Security Protocols

In order to enhance security in the IoT environment, various security protocols are available. They are listed below with their functions.

(i) **Security within IoT protocol layers:** It allows some security layers inbuilt with the TCP/IP layers. They are 802.15.42 security enabled bit in frame, Wireless HART-AES 128 encryption, 6 LOWPAN-IETF document for security, RPL-Security field in the header.

(ii) **TLS (transport layer security)/DTLS (datagram transport layer security):** It enables authentication, integrity and confidentiality at transport layer and provides signature, credentials and error handling.

(iii) **Ubiquitous green computing control network security:** It enhances communication by maintaining high quality, energy saving, secure mechanism, information protect, integrity confidentiality, authentication and access control.

(iv) **TCG (trusted computing group):** It gives availability, confidentiality, integrity, protection against malware infection.

(v) **OAuth2.0:** It enables 3rd party server to provide security. The main disadvantage is the credential leakage, risk in 3rd party server and it is rarely used in IoT.

(vi) SASL(simple authentication security lags): Supported by session layer protocols, and it is an application-specific authentication mechanism.

(vii) ACE: It is built on COAP and more suitable for IoT.

5.3 NETWORK TOPOLOGIES IN IoT

A network topology defines how the computers, printers and other devices are connected. It describes the layout of wires, devices and routing path on a network. When started the term Internet of Things, how the sensors, actuators and gateways communicate with one another in IoT are referred to as network topology. Decide which network would be best for the selected smart application. Before starting the design of the smart system, the designer should need to know the advantages and disadvantages of the network. Most commonly used topologies for data transmission are point-to-point, bus, star and mesh. Figure 5.1 gives the model of different topologies and those are briefly explained as follows:

- Mesh topology
- Star topology
- Point-to-point topology
- Bus topology

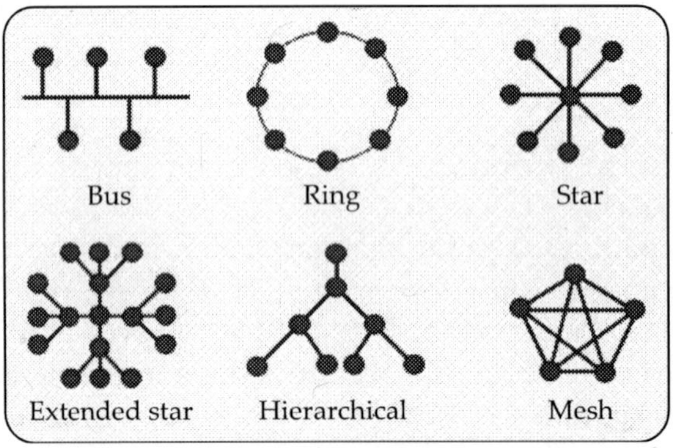

Fig. 5.1 Network topology

5.3.1 Mesh Topology

Mesh topology is a type of networking where all nodes cooperate to distribute data in a network. This topology is typically used for things like home automation, smart HVAC control, and smart buildings. The industry standards that rely on mesh network topology include ZigBee, Z-Wave, and Thread.

A mesh network consists of three types of nodes:

- A gateway node.
- Simple sensors nodes.

- Router nodes, (which are sensor nodes with repeater or routing capability).

Sensor/router nodes not only capture and disseminate their own data, but also serve as relays for other nodes.

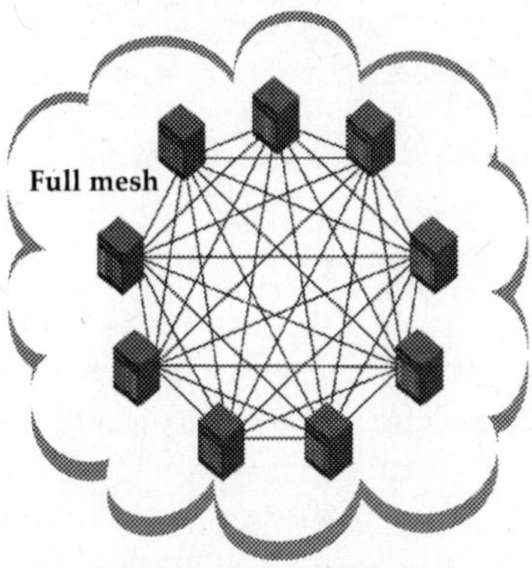

Fig. 5.2 Mesh network

Mesh networks (Figure 5.2) are deployed to make every node in the network to be in transmission range of at least one other sensor/router node. Data packets pass through multiple sensor nodes to reach the gateway node.

Mesh is a growing network topology commonly found in IoT deployment. Some of its advantages are listed below.

- Data can be transmitted from different devices simultaneously. This topology can withstand high traffic.
- Even if one of the components fails, there is always an alternative present. So data transfer doesn't get affected.
- Expansion and modification in topology can be done without disrupting other nodes.

5.3.2 Star Topology

The local area networks are constructed with a star topology. As shown in Figure 5.3, each host of the network is connected to a central hub with a point-to-point connection. Every node is indirectly connected to every other node with the help of the hub.

The switch is the server and the peripherals are the clients. The network does not necessarily have to resemble a star to be classified as a star network, but all the nodes on the network must be connected to one central device. All traffic that traverses the network passes through the central hub.

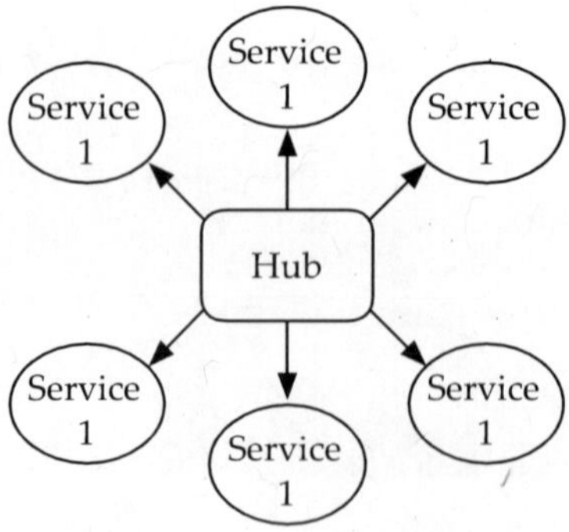

Fig. 5.3 Star topology

The advantage of star topology is the complexity in the network driven to a central node. So all the other nodes only need to communicate in their time or frequency slot.

The communication also depends on the wireless multiplexing like frequency-division multiple access (FDMA), time-division multiple access (TMA), or code-division multiple access (CDMA). The primary disadvantage of star topology is that the radio link between the gateway and the end node or terminal can be very long. It means that if a node is away from the gateway means it has to spend more energy for relaying a message.

But unlike a mesh node, it has to be constantly awake. Star nodes are able to rest between message transmissions. It helps to conserve the total amount of energy spends by each node.

5.3.3 Point-to-Point Topology

A point-to-point is shown in Figure 5.4. Network establishes a direct connection between two network nodes. Communication can take place only between these two nodes, or devices. An example of this type of network is a Bluetooth link between a cell phone and an earpiece. The advantages of point-to-point networking are its simplicity and low cost. The primary limitations spring from the one-to-one relationship that exists between two devices. The network cannot scale beyond these two nodes. Therefore, it is not a widely used topology for industrial IoT.

Fig. 5.4 Point-to-point topology

The advantage of point-to-point network topology is that it is much simpler than mesh or star. Because of this the topology has simply tunnels, a flow of data either in unidirectional or in bidirectional between two points occur.

The disadvantage is that point-to-point networks are not very useful for IoT. They are still used in some SCADA systems, traffic data systems, or in point-to-point broadcast systems (like police or fire radios), but it rarely makes sense in IoT to have a receiver talk to a single node instead of multiple nodes.

5.3.4 Bus Topology

The bus topology (Figure 5.5) is used in local area networks. Here each node is connected to the other node with the help of interface connectors (cable). This central cable is the backbone of the network and is known as the bus.

A signal from the source travels in both directions to all machines connected on the bus cable until it finds the intended recipient. If the machine address does not match the intended address for the data, the machine ignores the data.

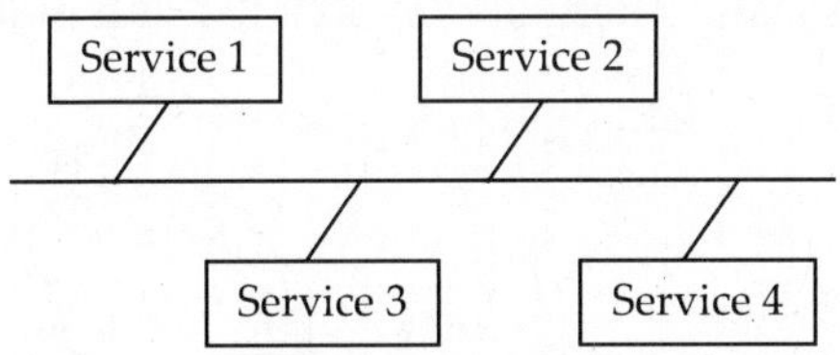

Fig. 5.5 Bus topology

Data sharing is also possible. If the data matches the machine address then the data is accepted. Because of only one wire is used in the bus topology, it is rather inexpensive to implement when compared to other topologies. However, the low cost of implementing the technology is offset by the high cost of managing the network.

5.4 ADDRESSING TECHNIQUES FOR IoT

In computer network, each system or host has a unique IP address for communication and also for making the network configuration. In wireless mode like laptop and printers the IP address is for sharing data. So IP addresses are needed to communicate over Internet in either wired or wireless mode. Now, the trends moving towards the IoT era, whether the smart system and its devices need IP address and if so who will provide? Addressing and identification is the main issues in the IoT network. With today's IoT environments and tomorrow's predictions, more conferences are conducted with a handful of experts observing their perspectives on how to allocate the addressing for IoT devices in the long term.

In this chapter let us discuss the different IP addressing strategies for nodes, gateways and various access network deployments in the Internet of things. In IoT, vast numbers of things are connected for the information sharing in the global network. IPV4 (Internet protocol version 4) has enough address space to allocate the devices connected in the network and also limited to local usage like point-to-point links. IPV4 supports addresses for laptops, smart phones, tablets and any other mobile devices. After the IoT era, IPV4 has less efficiency and moved to the IPV6 technologies to maintain and allocate IP address for IoT devices. The main drive is the large space addressing which are provided by IPV6. It also has auto configuration, security and flow identification that bring more advantages to IoT and also future technology.

The IoT network can handle connectivity management from a local network does to global network. The IPV6 does not ease the process to issue the scalability of the networking address and so has to configure some adaptive techniques to increase the efficiency of the IoT network. One of the most regular adaptive schemes in IPV6 addressing strategies is EUI-64 bits addressing scheme which produces static interface ID to assign address for wireless sensor network (WSN), IoT and mobile IPV6.

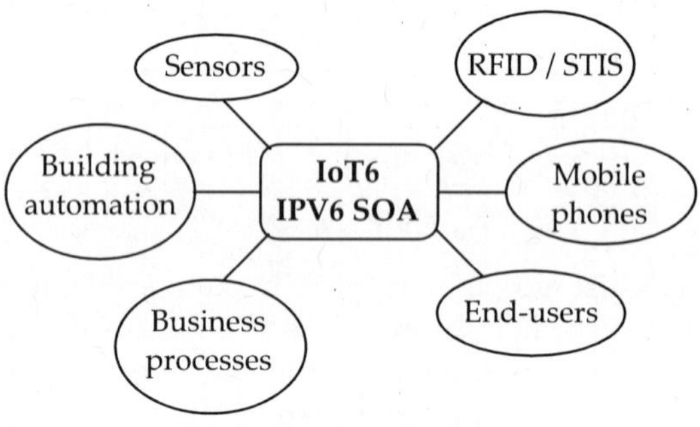

Fig. 5.6 Functions of IPV6

The IPV6 is capable of providing the address space for sensors, RFID, mobile phones, smart building, business process and end users to communicate with each other as shown in Figure 5.6. It enables interoperability to any heterogeneous global devices.

5.4.1 IPV6 Addressing Capabilities

Basic IoT elements are hardware like sensors, actuators and embedded devices, middleware or multiple platform accessible visualization and interpretation tools. Some IoT enabled technologies like RFID, WSN, addressing scheme, data storage and analytics, application domains influence IoT evolution. While making networks, some devices require unique network address.

Uriel kluk, CTO (chief technical officer) at mesh systems (Global IoT solutions) said that identifiers in IoT are the combination of both logical and physical address. For security purpose, company can allocate unique physical address to each device while designing. That physical address can be unique or it can be unique to the particular space also needs same kind of logical address. The combination of both addresses will give the real identity for each device. The same author warned that the devices with logical address create a lot of extra steps to make the devices work which causes delay in communication.

IoT consists of major components with more applications. Is it possible to provide unique address to all devices? Christian Legare, executive Vice President, CTO of Micrium Incorporated said that, "IoT predicts that it is not necessary to assign one global IP address to every device." Legare said "administrator shouldn't worry about how to allocate the unique IP address to each device. Every gateway or other aggregators are doing heavy networking. Each gateway has one unique public global address. Based on that, the IP addresses are allocated to connect to devices under this gateway. By that, the stress on address allocation is reduced and address allocated is based on the gateways are like that more number of nodes under one gateway. This is similar to, same STD code allocated to the trunk users with different telephone number."

Legare also said that many IoT devices don't have the same kind of common understanding that already exists in web technologies. It would take more than a controller to facilitate identification and information flow. Future adoption of protocols that either facilitates direct communication between devices and the rest of the network or the leverage gateway, controllers and similar devices may influence the types of structures to survive. So combination of intelligence at the edge and same authentication methods need to allocate the address.

5.4.2 IPV6 Features

IEEE 802.15.4g protocol support advanced metering infrastructure (AMI) for smart cities deployments.

- IPV6 specifications support for the operation and the deployment of IoT.
- IPV6 provides multicast, anycast, mobility support, and auto configuration and address space.
- Advance version of IPV6 are 6LOW PAN (wireless nets) COAP (transport with web services) and DTLS (secured datagrams).

These are the features of IoT addressing techniques of IPV6. Some of important parameters that need to know for IPV6 are given below.

1. *Adoption time*

The Internet protocol (IP) is the main requirement for any Internet connection. It is the addressing scheme for any data transfer over the Internet. This protocol provides very less or matter of time for adoption. The Google's figures revealing an IPV6 adoption rate is an exponential curve, doubling average of every nine months (Figure 5.7).

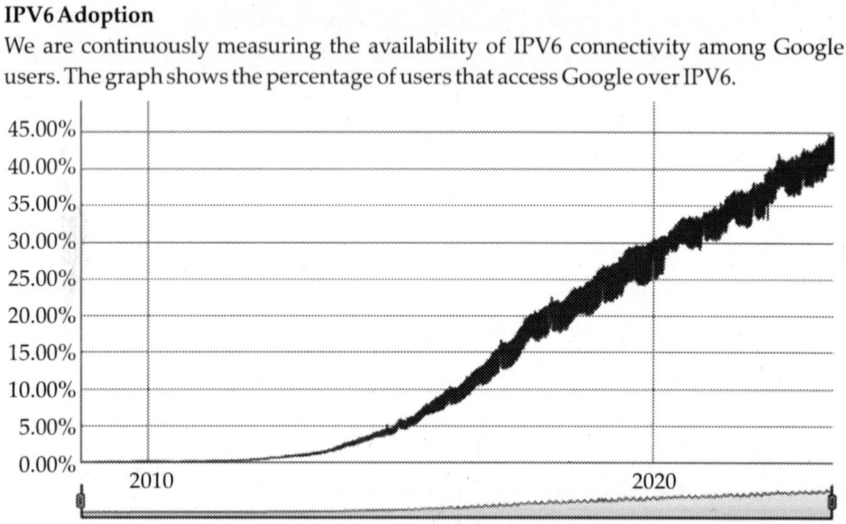

Fig. 5.7 IPV6 Adoption rate

(https://www.google.com/intl/en/ipv6/statistics.html)

2. *Scalability*

IPV6 is highly scalable and provides 128-bit address, that is, 2^128 = 3.4 × 1038 address. Billions address per square millimeter of the earth surface that will be used in this century. It is sufficient to address the present and future communication devices. It also provides new features enabling an easier configuration of devices.

3. *NAT (network address translation) barrier issues*

Due to the limitation of address space allocation in IPV4, the advanced version IPV6 adopted stopgap methods in unplanned expansion of network address translation. It can enable several users and devices to share the same public IP address. It also has two tradeoffs:

- The NAT users borrow and share IP addresses with others. So they do not have the own public IP address and they cannot be directly accessed from the Internet.

- It breaks the end-to-end connection and creates the issues in authentication process.

4. *Strong security supporter*

The IPV6 provides end-to-end connection with more distributed routes mechanism. It also supports for very large community of users and researchers move forward and improve the security features with IP security protocol (IPsec).

5. *Tiny OS or small stack*

The research community has developed a compressed version of IPV6 named 6LowPAN after being researched for many years. It is a simple mechanism to shorten the IPV6 address size for the constrained devices. The border routers can uncompress the address into regular size of IPV6 address. The researchers have developed more operating systems like Tiny OS and Contiki. TinyOS and Contiki are relatively small and support the protocol which suites and environments.

While the main IPV6 is rich in possible features, these reduced environments have often restricted carefully the features available in order to meet IoT needs while reducing the size of the underlying system and leaving more space for applications. For example, a basic Contiki system takes less than 20 kbyte, and even one supporting a full IPV6 stack and the other high level protocols including DTLS can probably fit into 70 kbyte. IPV6 provides for privacy by automatically randomizing the suffix of the IPV6 address to hide the MAC address or any serial number used as identifier when connecting to the Internet. This feature is made available on all operating systems automatically. The operating system and network stack (with security) could be made much more compact by providing more hardware support in the chipset (or a co-processor) which means increased hardware support.

6. *Enabling the extension to web of things*

IPV6 enables the extension of the Internet to any device and services by its large address space. The efficient address allocation by IPV6 has been successful verified by the deployment of sensors in smart building and smart cities and in the industries. The COAP protocol enables the constrained devices to behave as web services easily accessible and fully compliant with REST architecture.

7. *Mobility*

The IPV6 provides strong support and solutions to the mobility connection of end nodes and the mobility of the routes nodes of the network. (for moving nodes and moving routes).

8. ***Address self-configuration***

IPV6 provides an address self-configuration mechanism. The nodes can define their addresses in an autonomous way and reduces its configuration effort and cost. With an identifier-based system like handle, this technique can be combined with automated procedures to derive authentication tokens from the device, and have access control features added.

9. ***Multi-stake holder support***

The IPV6 provides support for end devices to have multiple addresses and an even more distributed routing mechanism than the IPV4 Internet. This allows different stakeholders to assign IoT end devices address with their own application and network practices. So it supports multiple stakeholders and they can deploy their own applications, sharing a common sensor/ actuator without impacting the technical expression or governance of the Internet.

5.4.3 IPV6 Architecture for IoT

Figure 5.8 describes the communication between sensor node, mobile node, Cloud computing node between Fog computing nodes with its three layers. The IPV6 supports transfer data between fog computing and mobile node, fog computing and Cloud computing. The μIPV6 is a special protocol to support communication between Fog computing and sensor

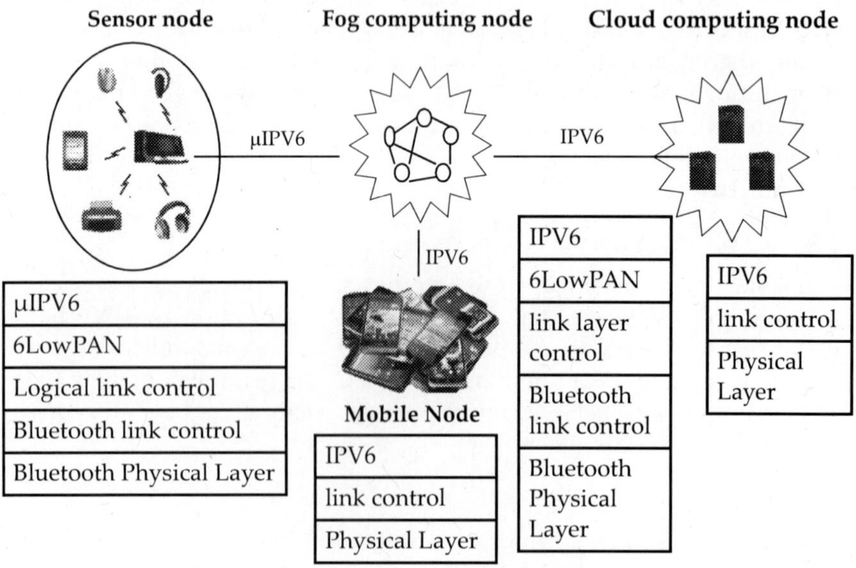

Fig. 5.8 IPV6 architecture of IoT

node. This is the additional feature in IPV6 which leads to support IoT based applications especially to provide more number of addresses to the sensors which is considered a node.

5.4.3.1 *IPV6 Header Structure*

IPv6 has fixed header size of 40 bytes long and contains all the information needed to carry the message from source to destination. Figure 5.9 gives IPV6 header structure. It consists of different fields which are described one by one.

Version (4 bits): The leftmost top four bits are used to represent the version of Internet protocol, i.e., 0110 (6).

Traffic Class (8 bits): The priority 8 bits followed with version is divided into two parts. Most significant 6 bits are used for type of service and the remaining used to assign priority to the packets.

VER (4 bits)	Priority or traffic class (8 bits)	Flow label (20 bits)	32 bits (4 bytes)
Payload length (16 bits)	Next header (8 bits)	Hop limit (8 bits)	32 bits (4 bytes)
Source IP address (128 bits)			128 bits (16 bytes)
Destination IP address (128 bits)			128 bits (16 bytes)
Payload extension header + data packet from the upper layer			

Fig. 5.9 Header format

Flow label (20 bits): This field is used to maintain the sequential flow of the packets belonging to a communication. The source labels the sequence which helps the router to identify the flow of the packet that moves towards the destination. This field helps to avoid re-ordering of data packets.

Payload length (16 bits): This field is used to tell the routers how much information this packet contains in its payload. Payload is composed of extension headers and upper layer data.

Next header (8 bits): This field is used to indicate either the type of extension header or it indicates the upper layer PDU.

Hop limit (8 bits): This field is used to stop packet to loop in the network infinitely. The value of hop limit field is decremented by 1 as it passes a link to the router/hop. When the field value reaches 0, the packet is discarded.

Source address (128 bits): This field indicates the address of the originator of the packet who is sending the message.

Destination address (128 bits): This field provides the address of intended recipient of the packet. IPV6 addresses use 128 bits to represent an address which includes bits to be used for site prefix, subnet and hosts address.

site prefix (48 bits)	subnet ID (16 bits)	interface ID (64 bits)

Fig. 5.10 128-bit address format

The 48 bits (1st three fields) is reserved as site prefix to describe the public topology allocated by an ISP (internet service provider) or RIR (regional internet registry). The next field is subnet ID with 16 bits to describe the private topology as shown in Figure 5.10. The rightmost four fields are reserved as interface ID or as a token. The interface ID is either automatically configured from the interface's MAC address or manually configured in EUI-64 format.

Extension headers: In IPV6, the fixed header contains only information which is necessary and avoides information which is either not required or is rarely used. When extension headers are used, IPV6 fixed header's next header field points to the first extension header as shown in Figure 5.11. If there is one more extension header, then first extension header's 'next-header' field point to the second one, and so on. The last extension header's 'next-header' field point to upper layer header. Thus, all extension headers are arranged one after another in a linked list type.

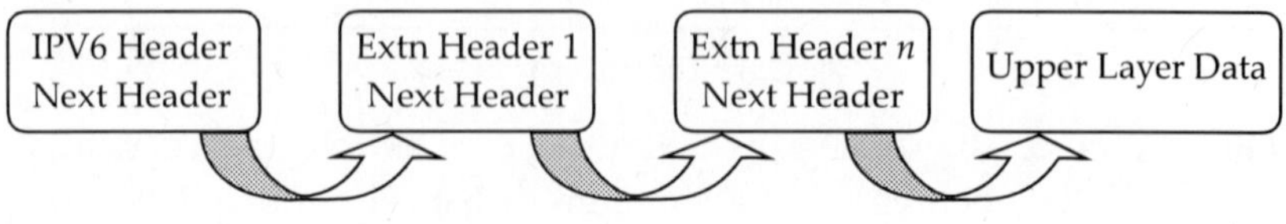

Fig. 5.11 Extension headers

5.4.3.2 *IPV6 Addressing Modes*

Addressing mode in IPV6 represents how to allocate the address to the host or device on the network. IPV6 provides different ways to address a single host or more number of hosts. It has three addressing methods:

- Unicast
- Multicast
- Anycast
- Unicast addressing mode

The unicast addressing mode is a unique address specification for the single host. The host interface is equipped with an IP address which is unique in that network segment. A network switch or router once receives a unicast IP packet, then it identifies a single host as shown in Figure 5.12, and sends out to that particular host through its outgoing interface connection.

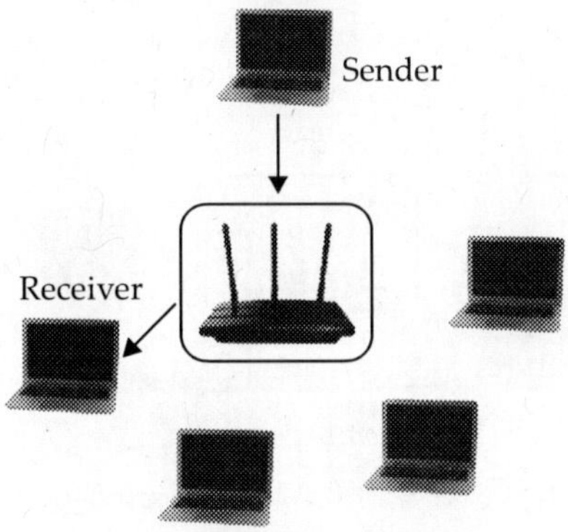

Fig. 5.12 Unicast mode

- **Multicast Addressing Mode:** The multicast addressing mode shown in Figure 5.13 is a special multicast address used to share the message to multiple hosts or to a group which acquires a multiple destination address. All hosts who have joined that group will receive the multicast packet and process it, while other hosts not interested in multicast packets ignore the multicast information.

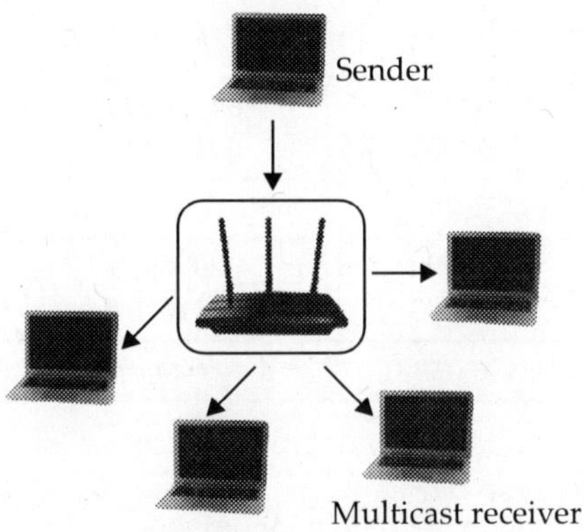

Fig. 5.13 Multicast mode

- **Anycast Address:** IPV6 has introduced a new type of addressing, which is called Anycast addressing. In this addressing mode, multiple interfaces (hosts) are assigned same Anycast IP address as in Figure 5.14. With the help of complex routing mechanism, that Unicast message is delivered to the host closest to the Sender, in terms of Routing cost. Any packet sent to Anycast address will be delivered to only one member interface (mostly physically closest to the sender).

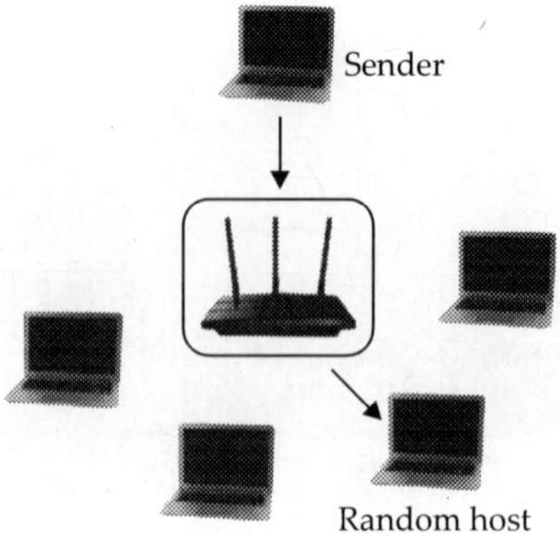

Fig. 5.14 Anycast address

5.4.3.3 IPV6 Address Format

IPv6 was developed by Internet Engineering Task Force (IETF). It specifies the Hexa colon representation of 128-bit addresses. This 128-bit source and destination address is divided as 8 groups of 16 bits or 2 bytes. Each block binary representation is converted into 4-digit hexadecimal numbers which are separated by colon (:) symbol. The binary format of 128-bit is given in Table 5.3.

Table 5.3 Binary format of IPV6128 bit

00010111 00001001	**00000000 00000000**	**00100000 00100000**	**00000000 01000101**
Block 1 (1709)	Block 2 (0000)	Block 3 (2020)	Block 4 (0045)
00000000 00000000	**00000000 00000000**	**00010000 00110101**	**01000111 00100110**
Block 5 (0000)	Block 6 (0000)	Block 2 (1035)	Block 2 (4726)

Each 4 bits are read as hexadecimal number of all blocks and is separated by colon as given below

1709:0000:2020:0045:0000:0000:1035:4726

These address specifications also are further reduced to increase the compatibility by discarding leading zeros or consecutive zeros.

- **Discard leading zeros:** The leading zeros in the block 4 (0045) are reduced as 45 by omitting two zeros. Then the address specified as

  1709:0000:2020:**45**:0000:0000:1035:4726

- **Discarding consecutive zeros:** The consecutive zeros in block 5 and block 6 are reduced as double colon '::' by omitting two block zeros. Then the address specified as

  1709:0000:2020:45:: 1035:4726

5.5 MIDDLEWARE IN IoT

5.5.1 What is Middleware?

Middleware is a software that serves as an interface between components of the IoT and making communication among different elements, without this middleware, communication would not happen. It can also be known as software glue because it provides a supporting layer between two layers for smooth communication.

Middleware enables connection and communication between various complex programs. It connects different components with different setup scenario. The trends of the Internet of Things are leads to make possible for just about anything (anything) to be connected and to communicate data over a network. Middleware is part of the architecture to enable the connectivity for huge numbers of different things by providing a connectivity layer. It provides the connectivity between the sensors and the application layers for remote access input.

5.5.2 Role of Middleware

Middleware in IoT is required for various reasons. The important role of middleware is given as follows:

- In general, it is difficult to define and enforce a common standard among all the different devices belonging to different domain in IoT. Middleware plays a major role in creating the interface between them.
- Middleware acts as a bond between the heterogeneous components together.
- Applications of different domains demand abstraction/adaptation layer.
- Middleware provides API (application programming interfacing) for physical layer communications, and required services to the applications, hiding all the details of diversity to perform communication operation between different environments.

In order to perform the above functions, there is a need of certain functional components in the middleware. It was briefly given in the following section. Middleware platforms can act as a key function of IoT by providing an infrastructure with a set of innovative services to seamlessly support the development of IoT applications. The IoT paradigm enables interconnections and information exchanges between the real-world objects and digital gadgets. IoT has been receiving more attention from the research community and industry, due to its potential of changing the world. Day by day it is providing a wide range of applications based on our daily lives. Although much progress has been made at the system and sensing level, there are several challenges to realize the vision of the IoT, in particular related to the support of applications development.

5.5.3 Functional Components of IoT Middleware

The need for various functional components in IoT middleware was clearly stated in the role of middleware. The important functional components are depicted in Figure 5.15.

The innermost circle shows the required functional blocks. The second circle encompasses further division of the functional blocks, and the outermost circle shows the important modules interacting with the various functional components, but not part of the middleware, for example, in the context processing, data storage and knowledge database. The functional block components are as follows:

- Interoperation
- Context detection
- Device discovery and management
- Security and privacy
- Managing data volume

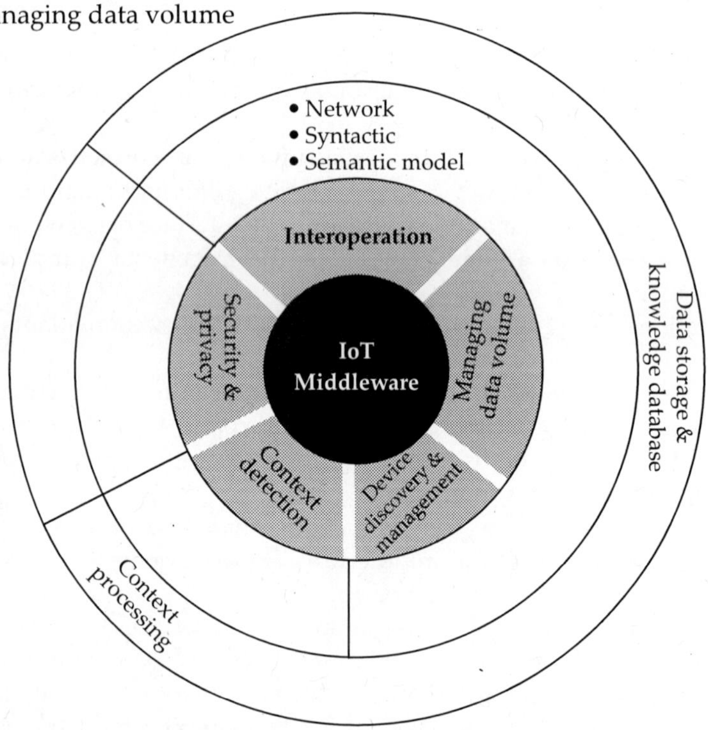

Fig. 5.15 Functional components of IoT-middleware

(https://www.researchgate.net/Fig./Functional-Components-of-IoT-Middleware-32_fig10_311797767)

5.5.3.1 *Interoperation*

Interoperation shares information and uses the same across different domains of applications using a variety of communication interfaces. It can be further classified under three different categories like network, syntactic and semantics.

- **Network interoperation:** Network interoperation defines protocols for exchanging information among the various things across different communication networks, without considering the content of information. It covers the basic connectivity issues in physical and data-link to network, transport, session and sometimes application layer of TCP/IP stack.
- **Syntactic interoperation:** Syntactic interoperation deals with the format and structure of the encoding information exchanged among things. It includes the presentation and application of TCP/IP stack.
- **Semantic interoperation:** Semantic interoperation defines the rules for understanding the meaning of the content of information, and creates a domain-specific information model, known as semantic model.

5.5.3.2 *Context Detection*

Context is responsible for characterizing the situation of an entity. Entity can be person, place, or object relevant to the interaction between a user and an application, including the user and applications themselves. IoT-middleware must be context aware for working into smart environments. Context awareness can be achieved by context detection and context processing.

Context detection collects data and identifies the factors that have significant impacts on the response and context detection is a core component of IoT-middleware. Context processing extracts the context data, processes it and performs or takes decision based on that. Both these functional modules are shown in Figure 5.15.

Context detection and context processing have been implemented in different ways in the existing IoT-middleware. Each service is supported by the middleware registered with a context-dependent identifier. The context-dependent identifier gets changed with the change in endpoint mobility.

5.5.3.3 *Device Discovery and Management*

- **Device Discovery:** Device discovery and management enables any device in the IoT network to detect all its neighboring devices and make its presence known to each neighbor in the network. Device ontology is used for storing information about the heterogeneous devices. From IoT perspective, these modules need to be reliable, fault-tolerant, adaptive and optimized for resource consumption.

Middleware described in various technologies tries to extend the syntactic interoperability to semantic interoperability in application layer. This is done by combining the use of ontologies with semantic web services. Semantic model driven architecture (semantic MDA) is introduced to facilitate application development and to promote semantic interoperability for services and devices.

Peer-to-Peer architecture is used in the system where devices support self-configurable services and scalability from tiny embedded devices to heterogeneous P2P networked systems. Using device ontology, sensor nodes are abstracted as virtual sensor node in some middleware. The virtual abstracted unit contains metadata used for identification and discovery of sensor nodes. It also contains information regarding the data stream, functional properties related to life cycle management, etc.

- **Management:** Middleware includes agents for managing the repository of roles and scenarios. Agents are able to monitor data coming from the adapter about states of the resource, and take decisions depending on the data content. Agents also facilitate discovery of other agents in the environment. This approach facilitates service discovery, FIPA communication protocols utilization, and integration/composition of services. Agent based layer includes an agent managing repository of roles and scenarios encoded in RDF-based semantic agent programming language (S-APL).

  S-APL is a hybrid of semantics specification languages, semantic reasoners, and agent programming languages. It integrates the semantic description of domain resources with the semantic prescription of behavior of agents in individual and collaborative level.

5.5.3.4 Security and Privacy

Security and privacy are responsible for confidentiality, authenticity, and non-repudiation. Security can be implemented in two ways: (i) secure high-level peer communication which enables higher layer to communicate among peers in a secure and abstract way, and (ii) secure topology management which deals with the authentication of new peers, permissions to access the network and protection of routing information exchanged in the network. Security has been considered in all the functional blocks of the existing IoT-middleware, from the user level application to the various parts of the functional blocks with a view to build a trustworthy design.

Communication and retrieval of trust information will be accomplished through corresponding agent-to-agent communication. Here agents represent communicating resources and required to be configured appropriately to handle all necessary trust management activities between the corresponding communication parties. Trust management procedures can be realized as a set of specific business scenarios in the form of agent configuration plans. Device authentication, integrity service and access control are other techniques deployed for IoT.

5.5.3.5 *Managing Data Volumes*

Managing data volumes is an integral part of IoT-middleware. It is believed that there will be trillions of objects which will be part of the IoT network and more number of data needs to be stored or exchanged among the objects. So, it is important to get novel methods to find, fetch, and transfer data. Mostly while managing data, the challenges involve in querying, indexing, process modeling, and transaction handling.

The storage manager (SM) module realizes the persistent storage and administration of information in the middleware. It can integrate any kind of storage into the middleware, address those storages as virtual devices and it stores the data as string. The main constraint of this middleware is that the storage device should also be running the same middleware. On the other hand, there is an existence of a RDBMS (relational data base management system). The information sharing module of the middleware is responsible for collecting, filtering, storing and extracting queried data from the database.

5.5.4 Forms and Services of Middleware

5.5.4.1 *Forms of Middleware*

As discussed above about middleware, it was clearly defined that middleware is a software which interfaces the client and the system. After the usage of middleware, programmers started to write machine code for specific platforms and designed in such a way that particular code can run on a specific system. In this situation the idea of a novel piece of middleware, the operating system, which lies between the hardware and the application to allow programmer to write code for an operating system instead of a specific hardware platform arose. Middleware takes various forms in high level categories based on their application and usage. Few forms of such application oriented middleware are discussed in the following sections.

5.5.4.2 *Transaction Processing Middleware (TP)*

Transaction processing (TP) monitor technology is shown in Figure 5.16. It involves in creating the software interfacing between the client and the transaction processing server. In order to perform various transaction operations, the client will perform the financial operation by using certain mobile gadgets or personal computers. The client starts doing the operations by logging in to certain apps (applications). Those apps act as a middleware to link the client and the transaction server.

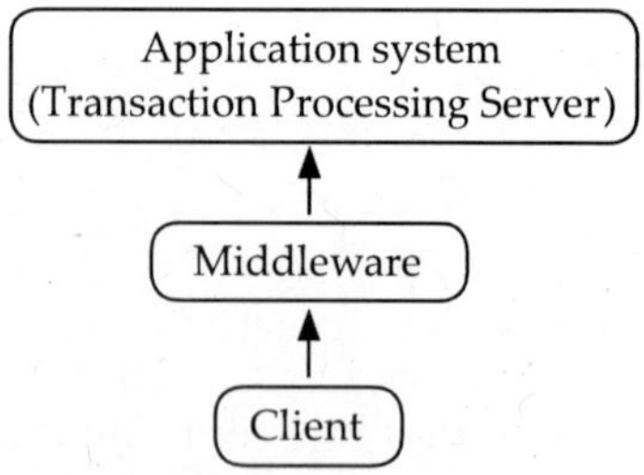

Fig. 5.16 TP monitor technology

TP monitor technology usually includes the ability for client to disconnect and reconnect seamlessly, restart failed processes to maintain availability. This type of middleware includes independence of the key layers. The client is independent and can be customized for specific platforms, the TP monitor is independent and it is limited to scalability. Examples of TP monitoring technology implementations are IBM's CICS TP monitor and BEA TUXEDO.

5.5.4.3 Message Oriented Middleware (MOM)

Message oriented middleware (MOM) in Figure 5.17 is a layer of software that presents on the operating system and provides communication services between applications either on the same machine or on other machines on possibly different platforms. The messages are generally asynchronous in nature but it could be synchronous. Synchronization can be done by middleware message queuing process. Then persistency is provided, when the message recipient is unavailable in a particular application which means the delivery is postponed until the recipient is available. This form of middleware provides interoperability, flexibility, and cross-platform portability services. Sometimes the asynchronous messaging service cause slowness due to overloading of the system which was the only drawback present in this form.

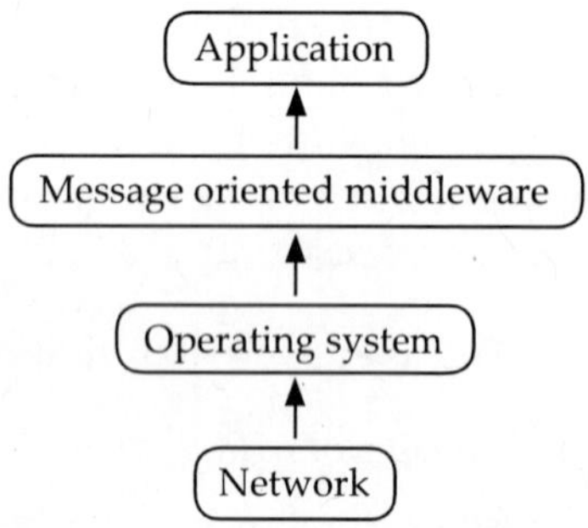

Fig. 5.17 Message oriented middleware

Examples of MOM systems are Oracle Advanced Queuing, IBM MQSeries, and Microsoft MSMQ.

5.5.4.4 Object Oriented Middleware (OOM)

Object oriented middleware (OOM) or object-request brokers (ORB) are middleware technologies that extend the object oriented development and communication paradigms to middleware. ORBs support interface definition languages, object communication mechanisms, and object activation and location mechanisms. ORBs facilitate locating objects and establishing communication with those objects in a manner similar to that of MOM technology. ORB interfaces and object communications usually appear as transparent extensions to the object oriented development environment. Objects that may be remotely located seem local to the objects communicating with them.

In addition, the interface is usually a natural extension of the object oriented development environment. Interface description languages employed by ORB systems provide a clear mechanism for providing contractual interfaces between objects, provide for rapid integration, and preserve implementation and interface separation.

Examples of object oriented middleware include OMG's CORBA, Microsoft's COM/DCOM, and Sun's Java RMI.

5.5.4.5 Database

Database connectivity issues are more complicated in the middleware. Mostly middleware products have been developed to provide API access to standard database interfaces in an effort to answer the call for simplified, robust database connectivity. This layer takes the form of development to connectivity tools and language extensions that facilitate the application to database communications. The advantages of this middleware come from its standard and simplified data-base connectivity. The disadvantages are that many solutions are not cross-platform, may not support advanced database features (especially proprietary ones), and may provide blocking, synchronous connections.

Examples of database middleware include Microsoft's ODBC, Sun's JDBC, and Rogue Wave Software.

5.5.4.6 Remote Procedure Call

A more discrete form of middleware lies in remote procedure call (RPC) technology. RPCs are stubs embedded in the client-server applications at compile time as shown in Figure 5.18. The stubs facilitate procedure calls between client and server. The advantages of RPCs include providing a level of consistency in procedure calls both locally and remotely, a natural extension for remotely calling

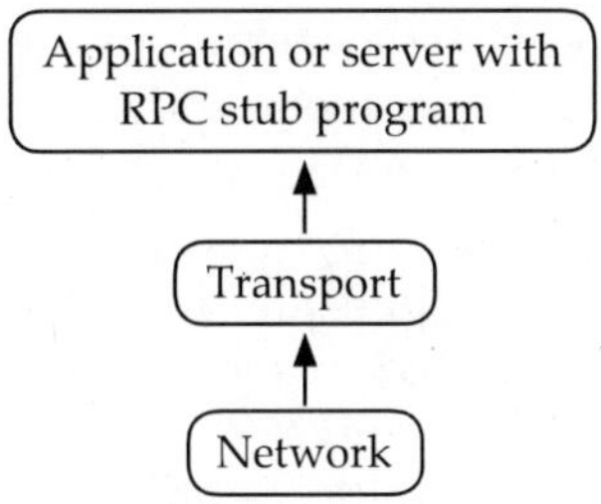

Fig. 5.18 Remote procedure call

server functions, and network transparency of the client-server locations. The disadvantages of RPCs are that it was mostly implemented using synchronous communications which forces a call waiting scenario and cause blockages in the flow of an application.

5.5.4.7 Web Services

Web services are one of the popular form of middleware because they bridge the interface gap between applications. A web service is simply a network accessible interface to an application using web based protocols (e.g., HTTP, XML). Typically, the interface is provided by ASCII text based XML (Extensible Markup Language) communicated between applications via HTTP (Hyper-Text Transport Protocol). XML merely specifies a format language. However, the true definition of the interface is molded by other emerging technologies such as SOAP (Simple Object Access Protocol) that add object oriented mechanisms to web services and WSDL (Web Services Definition Language), which provides communication descriptions between web service endpoints. Figure 5.19 illustrates how web services work.

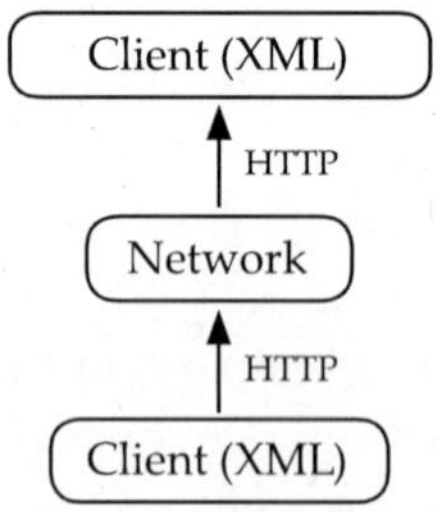

Fig. 5.19 Web services work

Examples of web service building tools and infrastructure support include Apple Computer's WebObjects, IBM's Web Sphere, Microsoft's .NET Framework, and Sun's Open Net Environment.

5.5.4.8 Agent Oriented

Agents are autonomous, intelligent software entities that have the ability to perceive their environment by reasoning themselves about what they observe, and act to change that environment to accomplish their goals. Agent technologies are used to solve financial management, military logistics, personal information management, and other problems. Since agents need to perceive and act on their environment, communicate with out-side entities and other agents, it may have platform and network mobility middleware components and multi-agent systems (MAS) to perform various agent based operations.

Examples of agent frameworks are HIVE [MGR+99] and CMU's RETSINA [SDP+96].

5.5.4.9 Frameworks

Frameworks may provide their own private middleware services and at the same time may utilize other common middleware services. Frameworks are targeted at specific domains and consist of software environments that provide an API, a user interface, and tools for application development and system management. For specific domains, frameworks may provide better solutions than general middleware and may be accurately called middleware themselves.

Examples of some frameworks include Lotus Notes, Microsoft Office, Transarc's Encina, Cognos, and HP's OpenView.

5.5.4.10 Middleware Services

Middleware has to provide services based on each and every application it is used for. The reliable service is the major thing that everyone is looking for. If the system is more reliable the service provided by the system seems to be excellent. So the services provided by the middleware mainly focus on the event based communication scheme. It can provide a more decoupled publish-subscribe communication method augmenting or replacing point-to-point communications for the application system. Services can add value and flexibility to middleware. Certain services make the system to be self configured which are in-built in most of the systems to provide better servicing facility.

Technologies such as ZEROCONF and Universal Plug and Play seek to alleviate the problem through zero configuration technologies that allow devices to self integrate and configure in their local environments. These technologies may become part of the middleware infrastructure and are already embedded in the operating systems of Apple and Microsoft. Zero configuration technologies are poised to replace name and service discovery middleware services in the application systems.

5.6 SUMMARY

General network topologies, layer-wise protocols of IoT, various addressing methods and data transmission header details of IoT provides were detailed for the readers understanding of IoT. In addition to this a middleware functions too discussed for end user application purpose. Various IoT informations collected from various researchers and IoT companies clubbed together in this chapter to provide a vast knowledge to get in to IoT.

REVIEW QUESTIONS

1. Which communication protocol is used in short distance wireless communication?
2. List the layers in IoT.
3. Describe the layer-wise classification of IoT.
4. What are the protocols used in transport layer of IoT?
5. What is the purpose of management protocols?
6. Why do we need security protocols?
7. What are the various types of network topologies in IoT?
8. What is IPV6?
9. Describe the architecture of IPV6 in detail.
10. What is middleware? Write its need in IoT.
11. What are the functional components of IoT middleware?
12. List out the types of services used in IoT middleware?

6

SECURITY CHALLENGES IN IoT

6.1 INTRODUCTION

In ancient days, the computer was used for document preparation and storage of data in a stand-alone system. It was protected from data loss and effects of virus. Then, it started to form the network to communicate with each other by LAN, MAN and WAN topology. Then the technology development pushes them to share the message via Internet by ISDN. After that, the networks are moved to wireless communication either by short range like Wi-Fi, Bluetooth or by long, range like satellite. In wireless mode, any number of users can enter and leave the network at any time to share their data in the form of text, image, video, etc.

People started to enjoy with this efficient wireless communication in smart ways. When they began to share their personal information bank transaction and account details, they have little hesitation to this way of unsecured gadgets and wireless communication. So, they have to believe the integrity and availability of the network facility. People expect secured transaction and communication of the network. Figure 6.1 gives the general model of a wired and wireless network.

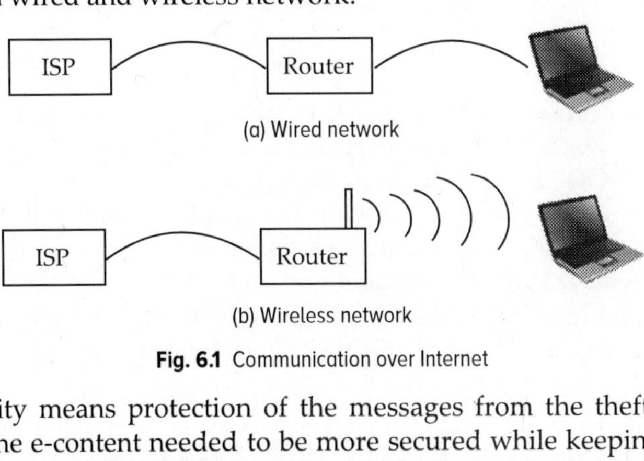

Fig. 6.1 Communication over Internet

Security means protection of the messages from the theft, lost and misuse. The e-content needed to be more secured while keeping the data in the system, communicating with other users and also bring in the secondary memory or removable disk. Security provides the protection of data and ensures that the integrity, availability and maintaining the performance of the network by protecting asserts from threats and thefts are taken.

Network is the connection of more system in either infrastructured or infrastructure-less network. Security should be provided for both networks between the end users to make secured network service. It preserves the integrity, availability and confidentiality among the information resource like hardware, software, firmware, data, and telecommunication. Security mechanism is applied to the network users to access their own information. The recent era begins with device-to-device and machine-to-machine communication by the support of Internet. This kind of advanced user friendly technology should be protected from attack, malfunction and theft.

For that every user has to use its own user id and password to access the information for authentication in network access. This kind of authentication verifies the user's information with the data stored in the database. This will protect from the intruder access to avoid malpractice or malfunction. If suppose any intruder came to know about the user id and password of the customer, then how to provide the security in data access? This leads to think about the safety of the data. The stored information and also during the communication the data should be processed in a secured manner. For that the encoded message could be stored in a database as the intruder couldn't decode the actual information quickly from the encoded information.

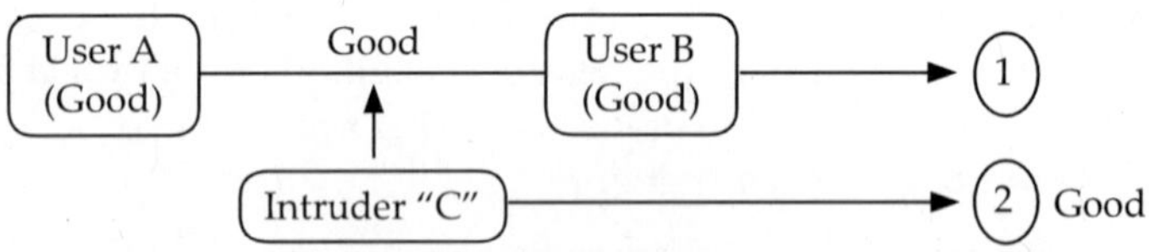

Fig. 6.2 Unsecured communication

Communication is performed between A and B users (Figure 6.2). The message "GOOD" is transmitted and the intruder C also gets the message "GOOD" by middleman attack. In case of encrypted or secured communication, the message "GOOD" is transmitted as "IQQF" from user A to user B and the intruder C also reads the communicated information as "IQQF" meaningless message and intruder couldn't understand the message immediately as shown in Figure 6.3. As the unsecured communication the message "GOOD" is termed as plain text nothing but

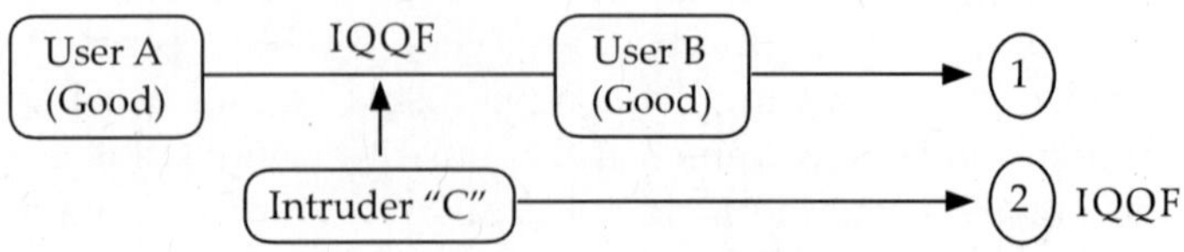

Fig. 6.3 Secured communication

the original information. In Figure 6.4 the encryption algorithm is used to change the plain text as cipher text which is stored or transmitted in a secured way.

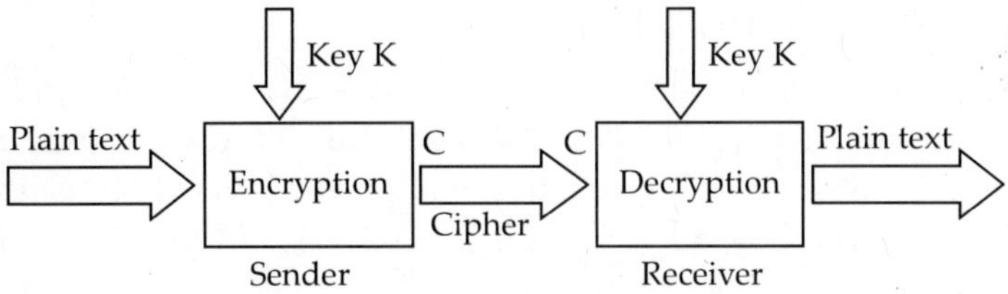

Fig. 6.4 Encryption and decryption process

At the receiver side, the received cipher text can be converted into the plain text by decryption algorithm to recover the original text. Some mathematical procedures have been used in the encryption algorithm for message with key. Simple addition and multiplication are used for encryption. For recovering the data additive inverse and multiplicative inverse are used between cipher text and key in the decryption algorithm.

The concept of encryption and decryption scheme is called cryptography. The role of cryptography is achieving security by encoding message to make them non-readable or non-understandable. Substitution and transposition techniques are used in the cryptography. The ciphering will be either block of information called *block cipher* or streaming of information called *stream cipher*. Stream is the bit-wise ciphering and block is the byte or word-wise ciphering.

Techniques used to decipher the message without any idea of encipher is called *cryptanalysis*. It means that the breaking of text or code for trying to find the key which was used for encryption. If the intruder can find the key, then easily decipher the cipher text or original text. The researchers suggested to design the irreversible keys or should take more time to analyze. So the complicated polynomials are suggested to generate the key for creating the cipher text. The study of cryptography and cryptanalysis is called *cryptology*. Same key is used for encryption and decryption, and then it is called *symmetric cipher* otherwise called *asymmetric cipher*. Now the protection is extended to Internet enabled devices which are connected to each other on wireless networks. It is an IoT security to provide safety for the IoT devices and its network against cyber crime.

The security vulnerability is the weakness that can be exploited to compromise the integrity or availability of connected IoT applications. Many of the IoT devices have a low level of concept, minimum memory and storage capabilities which limits the opportunities to implement the security on the devices. Even after implementing the security, it might be affected by different new attacks by emerging technology. So to detect and mitigate the vulnerabilities, the service providers, manufacturers and organization should watch consistently and audit device settings.

6.2 SECURITY MEASURES IN IoT

The security measures are taken as a precaution against theft and loss. Security is invisible and difficult to measure. But its functions are more visible and demonstrable. In vehicle area network more vehicles are connected and their systems relying more on complex networked information and protecting the information is a priority task. In a vehicle, information covers the details of the user, the information exchanged between electronic systems and even the software that is stored to make the systems work. If anyone attacks against your travel plan then what will happen? No one could imagine about the risk factors. In public bus access, anyone can hack and change the bus route. Cyber security simply means that the information is protected against criminal or unauthorized use and/or that measures are taken to achieve this. In order to analyze cyber security, one has to understand the threat and security measures which are a well known model for cyber security development. Automotive systems and related infrastructure must be protected against deliberate or accidental compromise of confidentiality, integrity or availability of the information that they store, process and communicate without hindering safety and functionality. It is important to understand each of these concepts because all risks, threats and vulnerabilities are measured for their potential capability to compromise one or all of these principles. So the security measures involve three basic concepts (Figure 6.5). They are confidentiality, integrity and availability (CIA)

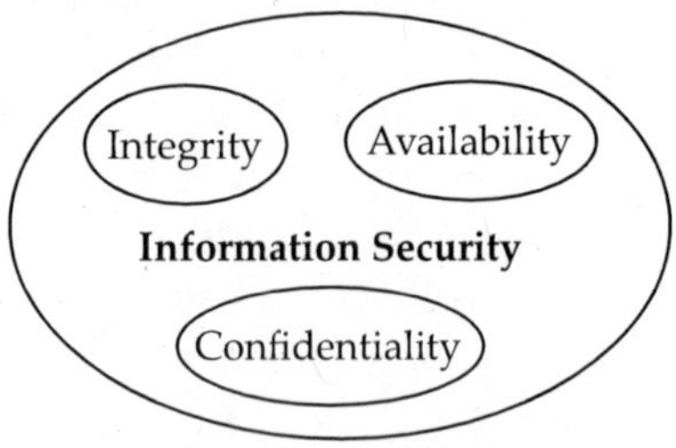

Fig. 6.5 Security measures

6.2.1 Confidentiality

Confidentiality measures to protect information from unauthorized access. As shown in Figure 6.6, it ensures no data exchanged between unauthorized users. Threat vectors include direct attacks such as stealing passwords and capturing network traffic, and more layered attacks such as social engineering and phishing. A few types of common accidental breaches include emailing sensitive information to the wrong recipient, publishing private data to public web servers, and leaving confidential information displayed on an unattended computer monitor. These types of attacks are passive attacks which will display the content without modifying.

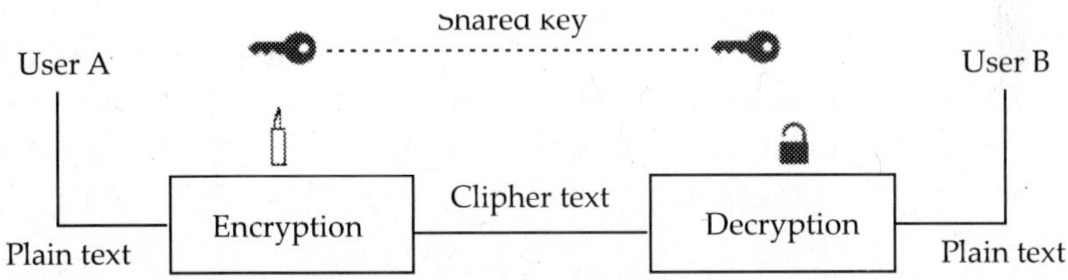

Fig. 6.6 Confidentiality

Consider an example of protecting patient's health information. Not only do patients expect and demand that healthcare providers to protect their privacy, there are strict regulations governing how healthcare organizations manage security. The Health Insurance Portability and Accountability Act (HIPAA) addresses the security including privacy protection, in handling of personal health information by insurers, providers and claims processors. HIPAA rules mandate administrative, physical and technical safeguards, and require organizations to conduct risk analysis.

Some of the possible attacks and the need of counter measures are discussed. The intruder can change time in smart watch, without user's knowledge is an active attack which is modifying the content and will cause vulnerabilities. The attacker may change the wake-up time or dialysis time and this can change the effect of vulnerability. There are many countermeasures that organizations put in place to ensure confidentiality. Passwords, access control lists and authentication procedures use software to control access to the resources. The device access control should be the user authentication process by making user id and strong password protection like e-mail access. At the same situation, we have to keep the information to be secured by applying encryption for information security like e- content bank statement and e-adhaar. Additional confidentiality countermeasures include administrative solutions such as policies and training, as well as physical controls that prevent people from accessing facilities and equipment.

To maintain confidentiality in automotive systems, data needs to be protected inside and outside the vehicle, while it is stored (data at rest), while it is transmitted (data in motion), and while it is being processed (data in use). Memory protection can be applied to data in use. Cryptography is excellent for protecting the confidentiality of data at rest and data in motion, but keep in mind that it imposes computational complexity and increases latency, so it should be used with caution in time-sensitive systems.

In IoT network, a user can be humans, machines and services, and internal objects (devices that are part of the network) and external objects (devices that are not part of the network). For example, it is crucial to make sure that sensors don't reveal the collected data to neighboring nodes. One more confidentiality issue that must be addressed is how the data will be managed. It is important for the users of IoT to be aware of the data management mechanisms that will be applied, the process or person responsible for the management, and to ensure that the data is protected throughout the process.

In IoT, the devices and sensors need to exchange some encryption materials to ensure confidentiality of the data. For this purpose, there needs to be a light-weight key management system for all frameworks that enable trust between different things, and can distribute keys by consuming devices minimum capabilities.

6.2.2 Integrity

The general model of integrity is given in Figure 6.7. The user A message create the digest using Hash algorithm and encrypted and then transmitted to the user B. Now user B also creates a digest from the received message which will be compared with received digest from user A. If both are same then the signature between the users are verified for the integrity. Integrity measures to protect information from unauthorized alteration. These measures provide assurance in the accuracy and completeness of data. It is the ability to ensure that a system and its data has not suffered unauthorized modification. Integrity protection protects not only data, but also operating systems, applications and hardware from being altered by unauthorized individuals. In automotive systems, CRC is known to

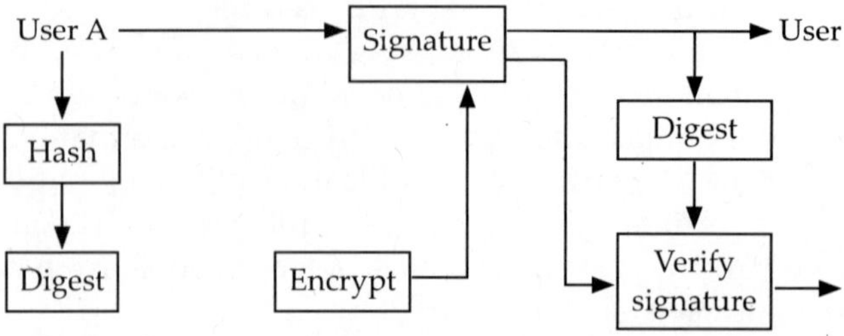

Fig. 6.7 Integrity

provide integrity protection against accidental or non-malicious errors; however, it is not suitable for protecting against intentional alteration of data. Hence, the sensitive data should include cryptographic checksums for verification of integrity. The need to protect information includes both data that is stored on systems and data that is transmitted between systems such as email. In maintaining integrity, it is not only necessary to control access at the system level, but also to further ensure that system users are only able to alter information that they are legitimately authorized to alter. As with confidentiality protection, the protection of data integrity extends beyond intentional breaches. Effective integrity countermeasures must also protect against unintentional alteration, such as user errors or data loss that is a result of a system malfunction by virus.

While all system owners require confidence in the integrity of their data, the finance industry has a particularly pointed need to ensure that transactions across its systems are secure from tampering. One of the most notorious financial data integrity breaches in recent times occurred in February 2016 when cyber thieves generated $1 billion in fraudulent withdrawals from the account of the central bank of Bangladesh at the Federal Reserve Bank of New York. The hackers executed an elaborate scheme that included obtaining the necessary credentials to initiate the withdrawals, along with infecting the banking system with malware that deleted the database records of the transfers and then suppressed the confirmation messages which would have alerted banking authorities to the fraud. After the scheme was discovered most of the transfers were either blocked or the funds recovered, but the thieves were still able to make off with more than $60 million.

There are many countermeasures that can be put in place to protect integrity. Access control and rigorous authentication can help to prevent authorized users from making unauthorized changes. Hash verifications and digital signatures are used to ensure that transactions are authentic and that files have not been modified or corrupted. Equally important to protecting data integrity are administrative controls such as separation of duties and training.

The IoT is based on exchanging data between many different devices, which is why it is very important to ensure the accuracy of the data. Incoming data from the right sender was as well as to ensure that the data was not tampered during the process of transmission due to intended or unintended interference. The integrity feature can be imposed by maintaining end-to-end security in IoT communication. The data traffic is managed by the use of firewalls and protocols, but it does not guarantee the security at endpoints because of the characteristic nature of low computational power at IoT nodes.

6.2.3 Availability

Availability measures to protect timely and uninterrupted access to the system. It should be guaranteed to use the systems, applications and data available to users when they need to access. The information should be available to the authorized user. The most important fundamental threats are non-malicious in nature and any hardware failures, unscheduled software downtime and network bandwidth issues. Malicious attacks include various forms of sabotage intended to cause harm to an organization by denying users access to the information system. It is also due to the denial of service to the user for accessing the information system.

The denial of service (DoS) is one kind of attack frequently used by hackers to disrupt web services. The attacker interrupts access to information, system, devices or other network resources. A denial-of-service in an internal vehicular network could result in an ECU (electronic control unit) not being able to access the information needed to operate and the ECU could become nonoperational or even worst it could bring the system to an unsafe state. To avoid availability problems, it is necessary to include redundancy paths and fail over strategies in the design stage, as well as to include intrusion prevention systems that can monitor network traffic pattern, determine if there is an anomaly and block network traffic when needed. In a DoS attack, the hackers flood a server with superfluous requests, overwhelming the server and degrading service for legitimate users. Recently, the service providers have developed sophisticated countermeasures for detecting and protecting against DoS attacks, but hackers also continue to gain in sophistication and such attacks remain an ongoing concern.

Availability countermeasures to protect system availability are as far ranging as the threats to availability. Systems that have a high requirement for continuous uptime should have significant hardware redundancy with backup servers and data storage immediately available. Countermeasures to protect against DoS attacks include firewalls and routers.

6.2.4 Need of Security Measures in IoT?

The following security measures are used in recent IoT development.

- **Trusted Platform Module (TPM):** The TPM is a dedicated microprocessor that integrates cryptographic keys into devices to uniquely identify and authenticate them. This module is mainly for providing authentication process.
- **Trusted Network Connect (TNC):** The TNC standards provide a way to check devices for malicious software or firmware whenever they try to access networks or other devices. It is a device level checking which

is to identify the attacks by malicious software or firmware during communication with new network.

- **Isolation:** If a device is infected with malware or other malicious programs, it needs to be quarantined. Immediately isolate the infected device and remediate infected devices with security software and protocols.
- **Layered security:** The layer of security can limit the damage a hacker can do once the device is hacked. A mandatory access control system limits access to certain functions or files on a device for a given user. Once it is identified, then the remaining layers will deny the access.
- **Data encryption:** The data encryption is very important. This should go without saying, but data needs to be encrypted when stored on a device or in transit. If anything happens, the intruder cannot access the data because of the hidden encrypted data.

6.3 SECURITY ISSUES IN IoT

The security in IoT means that the protection of IoT devices and communications from attack. The device production or security is easy one by keeping password. So the risks are less in device level. The IoT devices are used everywhere around us. From vehicles to home most of the automated systems are increasing and monitored or service connected to the Internet. IoT involves supporting Internet connectivity to the system of interrelated computing devices, mechanical and digital machines, objects, animals and people. Each one (thing) is operated with a unique identification and can provide the ability to automatically transfer data over a network. Allowing devices to connect to the Internet opens them up to a number of serious vulnerabilities if they are not properly protected.

The digital technology growth tremendously increases with advancements and modernization. In that the IoT techniques play a vital role in the smart environment. Successfully, the smartness and digitization are implemented in vehicles, bank, medical testing and treatment, agriculture field, environment science, etc. Now the customer started to think about the threats and attacks over the e-content and e-communication. For that, the security has to implement in device and connection. Antivirus software can be installed in the device and periodically update to avoid the attacks in device level.

6.3.1 Security Issues Level

Research started in IoT for providing security access between the devices. But till now we do not have universal IoT security solutions. It is an emerging and important field for conventional computing devices, the desktops, servers and smart phones run on a handful of platforms. The traditional network security is provided with an endpoint level and firewall. But same thing couldn't support for IoT network security. It is a bit more challenging than traditional network security because it has an endless number of communication protocols, standards, and device capabilities based on user requirements like dynamic network, all of which pose significant issues and increased complexity to develop a standard security solution. So instead of searching algorithm for a standard security suite, IoT security relies upon securing the parameter, the storage, the platform and the communication between two devices or the Internet gateway. Based on these, the IoT security has been classified as three levels:

- Network level security
- Device level security
- Communication level security

6.3.1.1 Network Level Security

Network security is the process of taking hardware and software preventive measures to protect the underlying network information from unauthorized access, misuse, malfunction, destination or improper disclosure, thereby creating a secure platform for computers, users and program to perform. In IoT network, the security is concentrated on the end point and firewall like existing network.

The communication in the IoT is different from that of the Internet because it is not restricted to machine to human. However, the feature of machine-to-machine communication in IoT introduces a security issue of compatibility. The heterogeneity of the network components makes it difficult to use the current network protocols and still produce efficient protection mechanisms. Protecting the network is important in the IoT but protecting the objects in the network is also equally important. Objects must have the ability to know the state of the network and the ability to protect themselves from any attacks against the network. This can be achieved by having good protocols as well as software that enable objects to respond to any situations and behaviors that can be considered abnormal or may affect their security.

- **Endpoint security:** The endpoint security which protects the enterprise networks when accessed via a computing device such as a tablet, smart phone or a laptop can protect IoT devices too. An IoT device creates a potential entry point for security threats and the endpoint security filters them out.
- **Network firewall:** A firewall is a network security system (Figure 6.8) that filters incoming and outgoing streams of network traffic based on a set of predefined security rules. A firewall creates a virtual barrier that separates trusted internal network from untrusted external network,

such as the Internet to which IoT devices and other mobile and computing devices are connected to it. Network firewalls can be based on either software or hardware. Most software security suites include a firewall with a predefined set of rules. However, software firewalls are easy to break into than hardware firewall.

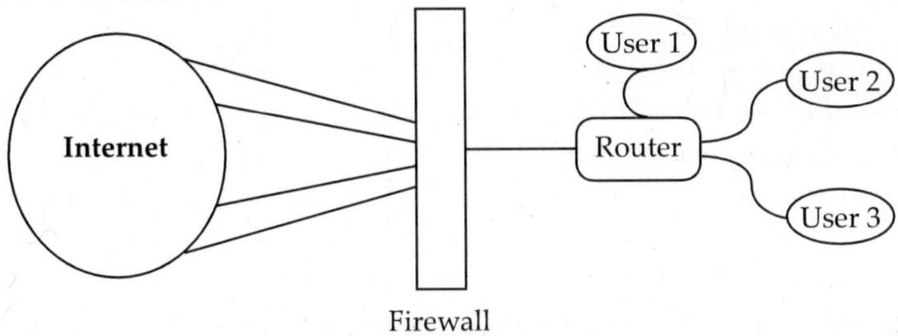

Fig. 6.8 Network firewall

6.3.1.2 Device Level Security

Device level security is to protect the device and its information from the attackers as shown in Figure 6.9. A server connected with smart bulbs, door lock, smart TV, IoT speakers which all are connected in wireless. Attacker can hack any one of the smart devices in the server control and take over the overall control. Consider the following scenarios to realize the effect of security attacks.

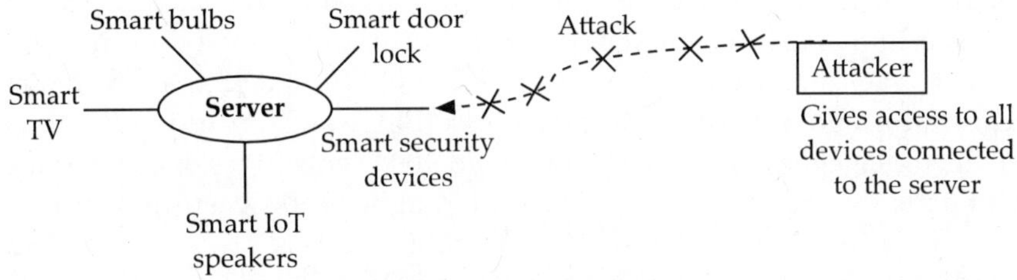

Fig. 6.9 Device level security

(i) Anyone might want to raise your thermostat's temperature to an unprecedented level to roast you alive, does that sound scary?
(ii) Or an unknown firm trying to get access to the personal data, generated by your IoT devices for commercial gain.
(iii) Covert operators trying to listen to you with the mic connected to your smart TV after gaining access to it or the computer network connected via an unscrupulous way.
(iv) Not to mention, government surveillance department has access to your home network.

So the security for device level is essential one. The users have to keep the device safe from the damage and have to keep the information safe by pattern to the device access. Generally, most of the risks are raised from the manufacturers due to weak, guessable, or hard coded setting of passwords, hardware issues, lack of a secure update mechanism, old and unpatched embedded operating systems and software and insecure data transfer and storage are some issues.

Recently, more number of new IoT devices come out almost daily, all with undiscovered vulnerabilities. The primary source of most IoT security issues is that manufacturer's do not spend enough time and resources on security. While listening music via the Bluetooth speaker, anyone can hack your speaker and play their songs without the concern of original owner. Washing machine time adjusted or the machine OFF by the neighbor during shortage of water. So the users have to provide device authentication by user id and password or biometric authentication.

This is precisely one of the biggest security issues with IoT. The lack of universal IoT security and standards leads manufacturers to create devices with poor security. Users are well expert to use Internet and how to avoid the unwanted pop ups, spam, phishing, e-mails, virus scans on their systems and how to use secured Wi-Fi network with strong password. People still lack how to protect the device and information in IoT technology.

One of the biggest IoT security risks and challenges is the user's ignorance and lack of awareness of the IoT functionality which put everybody into risk. A type of IoT security risk that is often overlooked is social engineering attacks. Instead of targeting devices, a hacker targets a human, using the IoT. Social engineering was used in the 2010 Stuxnet attack against a nuclear facility in Iran. The attack was directed to industrial programmable logic controllers (PLCs), which also fall into an IoT device category. The attack corrupted 1000 centrifuges and made the plant explode. It is believed that the internal network was isolated from the public network to avoid attacks, but all it took was a worker to plug a USB flash drive into one of the internal computers.

IoT security methods vary depending on your specific IoT application and your place in the IoT ecosystem. For example, IoT manufacturers from product makers to semiconductor companies should concentrate on building security from the start, building secure hardware with tamper proof and ensuring security upgrades by providing firmware updates/patches and performing dynamic testing while giving solution to the security issues. Manufacturer's should focus on secure software development and secure integration. For those deploying IoT systems, hardware security and authentication are critical measures. Likewise, for users keeping systems up-to-date, mitigating malware, auditing, protecting infrastructure and safeguarding credentials is a key feature to keep in mind for enhancing security at product level. Some of the important security features for device security are briefly explained in the following section.

- **Device Update Management:** IoT security risks occur due to insecure software or firmware. The manufacturer can sell a device with the latest software update almost inevitable to those new vulnerabilities during the manufacturing period. But based on the user application type and area, the device should be valuable and monitorable. The periodic updates of monitored values are needed by every user. Updates are critical for maintaining security on IoT devices. They should be updated before new vulnerabilities are discovered as shown in Figure 6.10. After updating the version, attacker can't attack the device. Still, as compared with smart phones or computers that get automatic updates, some IoT devices continue being used without the necessary updates. Updates should be shorter in time and smaller in size with the existing one. Sometimes the updated software is more difficult to use than existing. That's why the users hesitate or deny updating.

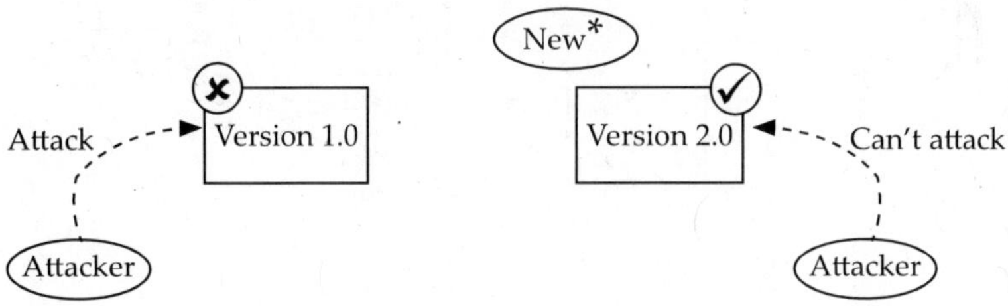

Fig. 6.10 Device update management

- **Physical Tampering:** Figure 6.11 shows the physical tampering which is to protect the device from the damage. The lack of physical hardening can also cause IoT security issues. Some of the IoT devices can operate autonomously without any intervention from a user, they need to be physically secured from outer threats. Sometimes, these devices can be located in remote locations for long stretches of time, and they could be physically tampered with, for example, using a USB flash drive with malware. Any tempering happens by the stranger. Then we have to send report message to the controller. Ensuring the physical security of an IoT device to build a secure sensors and transmitters with low-cost devices is a challenging task for manufacturers. Users are also responsible for keeping IoT devices in secured manner. The toughened, tamper-resistant setting, works as a trust anchor to guard the cryptographic substructure.

Fig. 6.11 Physical tampering

- **Botnet Attacks:** A hacker creates an army of bots by infecting them with malware and directs them to send thousands of requests per second to bring down the target is a botnet attack as shown in Figure 6.12. Much of the uproar about IoT security began after the Mirai bot attack in 2016. Multiple DDoS (distributed denial of service) attacks using hundreds of thousands of IP cameras, NAS, and home routers were infected and directed to bring down the DNS that provided services to platforms like GitHub, Twitter, Reddit, Netflix, and Airbnb. The problem in the IoT devices is highly vulnerable to malware attacks.

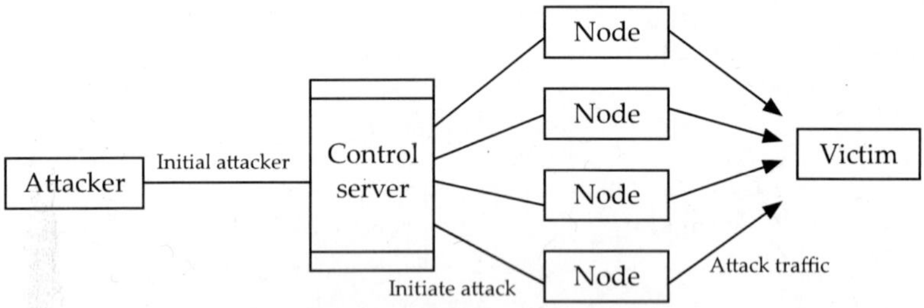

Fig. 6.12 Botnet attacks

- **Espionage and Eavesdropping:** If hackers take over surveillance at a location by infecting IoT devices, spying might not be the only option because they perform such attacks to demand ransom money. Thus, invading privacy is another prominent IoT security issue. Spying and intruding through IoT devices is a real problem, as lot of different sensitive data may be compromised and used against its owner. In Enthiran 2.0 film, the software hacked by the infected robot. Typing password is in a hidden way only but the number of dots is viewable. So anyone can guess the password by the number of variables used as pin. If the password has two dots with number means that the key will be found by 100 possibilities. If it was three dots then extended up to 1000 possibilities like that the complexity will increase with the length of password character. So the user should select more than six characters with alphanumeric which is more complex to find the pin. Figure 6.13 shows the eavesdropping attack between user A and user B.

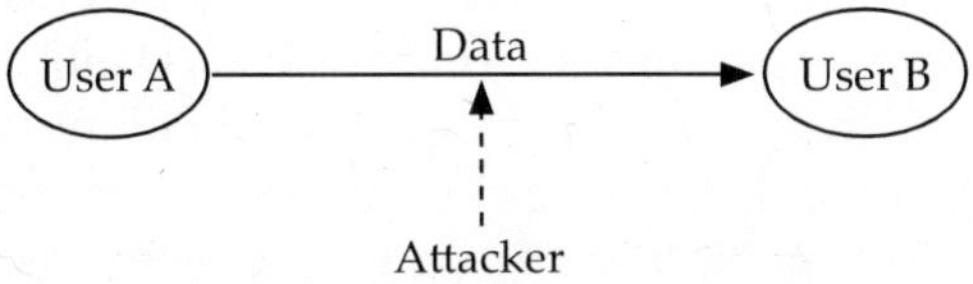

Fig. 6.13 Espionage and eavesdropping

Some countries have started to banning specific IoT devices with security problems. For example, the interactive IoT doll with a Bluetooth pin, which gave access to the toy's microphone and speaker to anyone within the 25-30 meter radius. The doll was labeled as an espionage device and was banned in Germany.

- **Hijacking IoT Devices:** Ransomware, as shown in Figure 6.14, is one of the malware which does not destroy the sensitive files but blocks access to them by way of encryption. Then, the hacker who infected the device will demand a ransom fee for the decryption key unlocking the files. Ransomware is evolving, and IoT devices with poor security can become targets as well. Just before the President Trump's inauguration speech, about 70% of the Washington DC surveillance cameras were infected with ransomware, leaving the police without the ability to record for several days. Still, wearables, healthcare gadgets, smart homes, and other smart devices and ecosystems might be at risk in the future. Because of this, malware couldn't lockdown the entire information of IoT storage. It also stored in Cloud if lockdown of the entire system functionality occur. The question mark raises how to protect the device? During the election time speakers of TV program will be hiden by either power down or software attack. Imagine that your vehicle will not start unless you pay a ransom fee or your house (smart lock) is locked down, with the thermostat set to the maximum, etc.

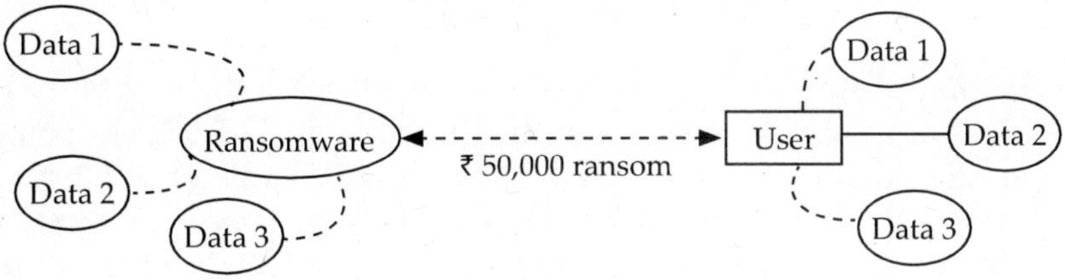

Fig. 6.14 Hijacking IoT devices

- **Data Integrity:** With IoT, data is always on the move. It is being transmitted, stored, and processed. The IoT devices extract and collect information from the external environment like a smart thermostat, HVAC, TV, medical devices. The devices send the collected data to the

Cloud without any encryption. As a result, a hacker can gain access to a medical IoT device, gaining control over it and being able to alter the data it collects. Consider some example for data integrity issues (i) A hacker controlled medical IoT device can be used to send false signals, which in turn, can make health practitioners take actions that may damage the health of their patients. (ii) A hacked medical IoT device can report a fully charged battery to the maintenance station while in reality the battery is about to die. Worse, there are risks of IoT security in healthcare devices like pacemakers or the ones making the insulin shots. The vulnerabilities found on St. Jude Medical's implantable cardiac gave access to hackers, enabling them to alter the pacing or shocks, or even worse, deplete the battery.

- **Rogue IoT Devices:** A rogue device replaces an original one or integrates as a member of a group to collect or alter sensitive information as shown in Figure 6.15. The user ID4 was hacked and denied its service and the hacker acts as Used ID4. These devices break the network perimeter. But rogue devices or counterfeit malicious IoT devices that are installed in secured networks without authorization.

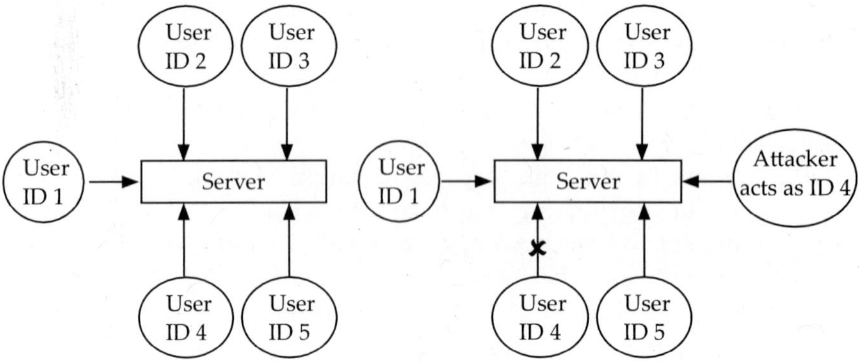

Fig. 6.15 Rogue IoT devices

Example of rouge IoT devices can take the form of the Raspberry Pi, or Wi-Fi Pineapple. These can be turned into a rogue AP (access point), thermostat, video camera, or MITM (man in the middle) and intercept incoming data communications unknown to users. Other variations of rogue devices may also emerge in the future. After the Mirai attack, people realized that any device connected to the Internet is a potential ally for an army of bots. But that was only the beginning. The IoT and security are still not found in the same place. There are still many risks and security challenges of IoT. International organizations and governments need to create universal IoT standards to control the security in cities, homes, locations like nuclear plants, the manufacturing process, and other areas and locations. If we want our devices smart, we need them to be secure as well.

A key manager authenticates IoT devices and monitors data exchanges taking place on cellular and non-cellular networks, such as LoRa, avoiding unsanctioned devices from connecting to the computer network. It makes way for a robust digital security over a trusted instrument of remote credential initiation, secure key provisioning, and application lifecycle management.

6.3.1.3 *Communication Level Security*

The IoT connects different entities with different capabilities, complexity, and different vendors. The devices even have different dates and release versions, use different technical interfaces. The IoT aims at connecting device to device, human to device, and human to human, thus it provides connection between heterogeneous things and networks. One more challenge that must be considered in IoT is that the environment is always changing (dynamics), at one time a device might be connected to a completely different set of devices than in another time. And to ensure security, optimal cryptography system is needed with an adequate key management and security protocols.

A key manager authenticates IoT devices and monitors data exchanges taking place on cellular and non-cellular networks, such as LoRa, avoiding unsanctioned devices from connecting to the computer network. It consists of four processes. They are

(a) End-to-end encryption
(b) Machine to machine authentication
(c) API security
(d) Cloud security

The communication security level between devices has been modeled as in Figure 6.16. It describes that (1, 2) shows the end-to-end encryption and machine authentication between the two devices, (3) indicates application interface between device and Cloud, and (4) provides security during information between transmitter and receiver.

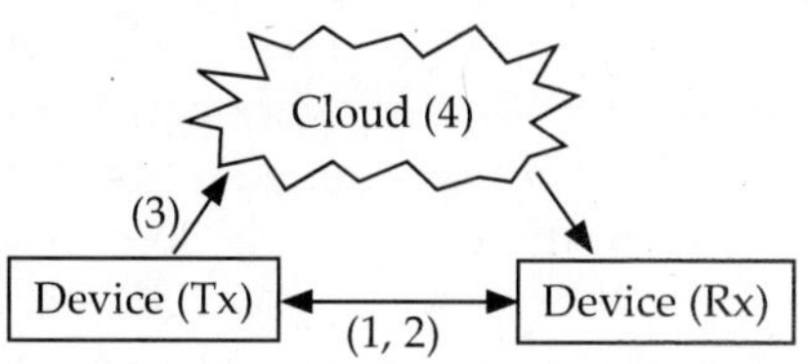

Fig. 6.16 Communication between device via Cloud.

(a) **End-to-end encryption:** It is to provide the end-to-end encryption between the devices while communicating with each other. The storage data should be encrypted and while starting to communicate, these encrypted messages only transfer to the receiving device via a secured channel as shown in Figure 6.17. If any intruder tries to get the data as middleman attack, he cannot quickly understand the encrypted word and also require more time to decrypt. The problem is in extensive array of IoT devices and hardware profiles that limits the capability to have usual encryption processes and conventions.

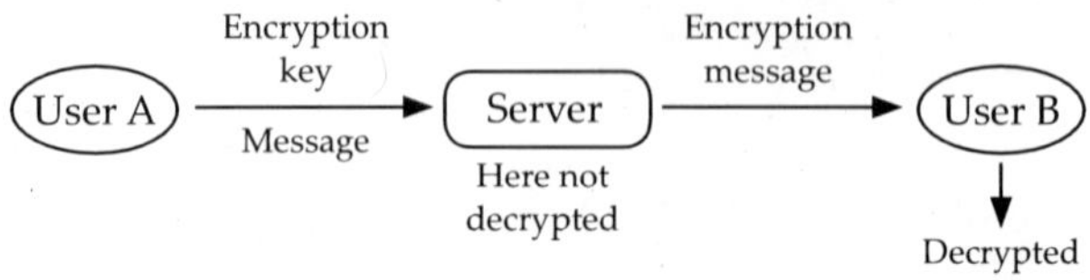

Fig. 6.17 End-to-end encryption

(b) **Machine-to-machine authentication:** For providing secure machine-to-machine communication, the authentication should be incorporated by the authentication process to validate the machines before transferring the information as shown in Figure 6.18. Many IoT authentication setups do not involve a person but machines on both sides making way for authentication methods that are machine-to-machine without the need of human intervention. More algorithms were developed to support the authentication process by the digital signature and digital certificate.

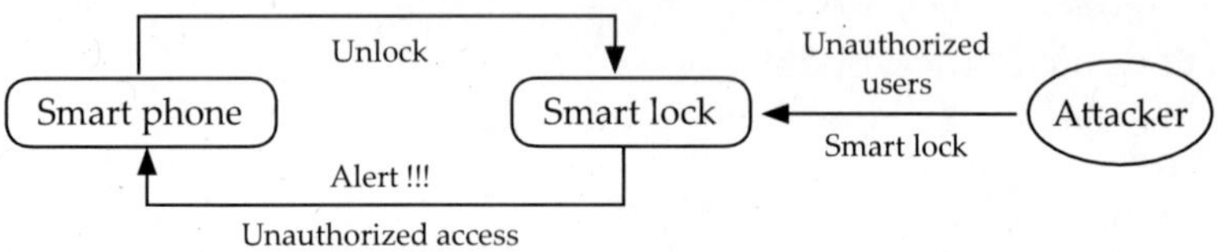

Fig. 6.18 Machine-to-machine authentication

(c) **API security:** Providing the capability to validate and sanction the data movement amongst IoT devices, backend schemes, and apps employ standard RESTful APIs. API security is critical for defending the integrity of data transiting between IoT devices and backend software infra to make sure that only sanctioned devices, certified developers, and trustworthy apps are collaborating with APIs as well as spotting possible threats and attacks against particular APIs.

(d) **Cloud security:** The IoT-based smart devices are connected through the Cloud to share the information. Cloud based environment will support for enormous amount of data storage. While sharing the data from device to Cloud and Cloud to device need to take care of avoiding data lost threats and attacks. A solution for data encryption and Cloud security offers an all inclusive portfolio for Cloud ecosystems and SMEs (small to medium sized enterprise) to protect their enterprise and Cloud assets. A Cloud based licensing and entitlement solution aids fast moving technology firm's leverage on the full potential of the Cloud environment, making sure their intellectual property is safe.

6.4 SECURITY CHALLENGES IN IoT

The main challenges of IoT can be broadly divided into two classes, i.e., technological challenges and security challenges. The technological challenges arise due to the heterogeneous and ubiquitous nature of IoT devices, while the security challenges are related to the principles and functionalities that should be enforced to achieve a secure network. Technological challenges are typically related to wireless technologies, scalability, energy, and distributed nature, while security challenges require the ability to ensure security by authentication, confidentiality, end-to-end security, integrity, etc. Security should be enforced in IoT throughout the development and operational life cycle of all IoT devices and hubs. The following are security precautions:

- The software running on all IoT devices should be authorized.
- When an IoT device is turned on, it should first authenticate itself into the network before collecting or sending data.
- Since the IoT devices have limited computation and memory capabilities, firewalling is necessary in IoT network to filter packets directed to the devices.
- The updates and patches on the device should be installed in a way that additional bandwidth is not consumed.

6.4.1 Technical Challenges

The security principles in technical challenges should be enforced to achieve a secure communication framework for the people, software, processes, and things. They are classified as CIA as discussed in security measures.

(a) **Confidentiality:** It is very important to ensure that the data is secure and only available to authorized users.

(b) **Integrity:** The integrity feature can be imposed by maintaining end-to-end security in IoT communication. The data traffic is managed by the use of firewalls and protocols, but it does not guarantee the security at endpoints because of the characteristic nature of low computational power at IoT nodes.

(c) Availability: The vision of IoT is to connect as many smart devices as possible. The users of the IoT should have all the data available whenever they need it. However, data is not the only component that is used in the IoT. Devices and services must also be reachable and available when needed in a timely fashion in order to achieve the expectations of IoT.

(d) Authentication: Each object in the IoT must be able to clearly identify and authenticate other objects. However, this process can be challenging because of the nature of the IoT. Many entities are involved (devices, people, services, service providers and processing units) and one other thing is that sometimes objects may need to interact with others for the first time (objects they do not know). Because of all this, a mechanism to mutually authenticate entities in every interaction in the IoT is needed. In 2011, some researchers presented a mutual authentication scheme for IoT between platforms and terminal nodes. The scheme is based on hashing and feature extraction. The feature extraction was combined with the hash function to avoid any collision attacks. This scheme actually provides a good solution for authentication in IoT.

6.4.2 Security Challenges in Each Layer of IoT

Each IoT layer is susceptible to security threats and attacks. These can be active, or passive, and can originate from external sources or internal network owing to an attack by the insider. An active attack directly stops the service while the passive kind monitors IoT network information without hindering its service. At each layer, IoT devices and services are susceptible to denial of service (DoS) attacks, which make the device, resource or network unavailable to authorized users.

6.4.2.1 Perception Layer

There are three security issues in IoT perception layer. First is the strength of wireless signals. Mostly the signals are transmitted between sensor nodes of IoT using wireless technologies whose efficiency can be compromised by disturbing waves. Secondly, the sensor node in IoT devices can be intercepted not only by the owner but also by the attackers because the IoT nodes usually operate in external and outdoor environments, leading to physical attacks on IoT sensors and devices in which an attacker can tamper with the hardware components of the device. Third is the inherent nature of network topology which is dynamic as the IoT nodes are often moved around different places. The IoT perception layer mostly consists of sensors and RFIDs, due to which their storage capacity, power consumption, and computation capability are limited making them susceptible to many kinds of threats and attacks.

The confidentiality of this layer can easily be exploited by replay attack which can be made by spoofing, altering or replaying the identity information in one of the IoT devices. Another way the attacker might gain the encryption key is by analyzing the required time to perform the encryption known as timing attack. Another confidentiality threatening attack is when the attacker takes over the node and captures all information and data which is basically node capture attack. Attacker can add another node to the network that threatens the integrity of the data in this layer by sending malicious data. This can also lead to a DoS attack, by consuming the energy of the nodes in the system and depriving it of the sleep mode that the nodes use to save the energy.

Security issues at perception layer can be addressed by using encryption (which can be point-to-point or end-to-end), authentication (to verify true identity of sender) and access control. Most of the devices involved in the perception layer of the IoT are RFID and sensors. Such devices have extremely limited computational capability, which makes it difficult to apply any cryptography algorithms to ensure the security of the network.

6.4.2.2 *Network Layer*

The network layer of IoT is also susceptible to DoS attacks. Apart from the DoS attacks, the adversary can also attack the confidentiality and privacy at network layer by traffic analysis, eavesdropping, and passive monitoring. These attacks have a high likelihood of occurrence because of the remote access mechanisms and data exchange of devices. The network layer is highly susceptible to man-in-the-middle attack, which can be followed by eavesdropping. If the keying material of the devices is eavesdropped, the secure communication channel will be completely compromised. The key exchange mechanism in IoT must be secured enough to prevent any intruder from eavesdropping, and then committing identity theft.

6.4.2.3 *Application Layer*

Since the IoT still does not have global policies and standards that govern the interaction and the development of applications, there are many issues related to application security. The large amounts of connected devices that share data will cause large overhead on applications that analyze the data, which can have big impact on the availability of the services.

6.4.2.4 *Additional layer (Session Layer)*

As per most of the researchers, the three-layer architecture of IoT does not accommodate the opening, closing, and managing a session between two things. So, there is a need for protocols which can address these issues and can ease the communication between devices. An abstract session layer should be accommodated as an additional layer in IoT architecture which can specifically manage the connections, protocols, and sessions between communicating heterogeneous devices. Table 6.1. describes various IoT layers and their security threats.

Table 6.1. IoT layers and its security threats

IoT layers	Security threats
Perception layer	Data security
Network layer	DoS
Application layer	CIA
Session layer	Heterogeneous support

IoT requires security measures at all three layers at physical/perception layer for data gathering, at network layer for routing and transmission, and at application layer to maintain confidentiality, authentication, and integrity. In the following section, the state-of-the-art security measures that address the specific features and security goals of IoT are discussed.

6.4.3 Authentication Measures

Researcher Wen et al. proposed for ID authentication technique at sensor nodes of IoT for providing authentication support in IoT network. It uses a one-time-one-cipher method based on request-reply mechanism. This dynamic variable cipher is implemented by using a pre-shared matrix between the communicating parties. The parties can generate a random coordinate which will serve as the key coordinate. Key coordinate is the thing which actually gets transferred between two parties, not the key itself. The key, i.e., password, is then generated from this coordinate. All the messages are sent by encrypting them with the key, along with key coordinate, device ID, and time stamp. The two devices communicate by validating timestamps, and thus they can cancel the session based on it. This cipher can be used where securing IoT is not very sensitive and crucial because key can be repeated for different coordinates. If key coordinate is changed regularly, security can be optimized for that particular IoT framework. The installation of pre-shared matrix needs to be secure for this work to be implemented for a large number of IoT devices.

Creating correct access controls is as important as authentication for security and these two functionalities will go hand in hand in the securing IoT. To address these functionalities, researchers Mahalle et al., presented an Identity Authentication and Capability based Access Control (IACAC) for the IoT. This research attempts to fill the gap for an integrated protocol with both authentication and access control capabilities to achieve mutual identity establishment in IoT. It prevents man in the middle attacks by using a timestamp in the authentication message between the devices, which serves as the message authentication code (MAC). The scheme as shown in Figure 6.19 works in three stages; first is a secret key generated based on elliptical curve cryptography Diffie-Hellman algorithm (ECCDH), then identity establishment is made by one-way and mutual authentication protocols and lastly access control is implemented by algorithm. This random number is the result of hashing device ID with access rights. The IACAC model does not completely prevent DoS attacks. However, it minimizes it since access of resource is granted to only one ID at a time.

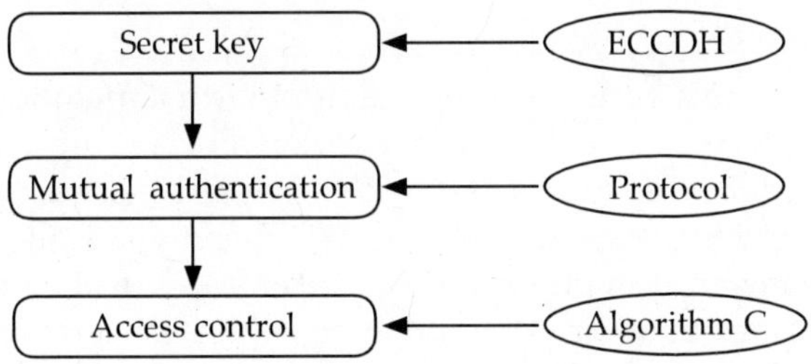

Fig. 6.19 Authentication measures

6.4.4 Trust Establishment

Devices in IoT can physically move from one owner to another. Trust should be established between both owners to enable a smooth transition between the IoT device with respect to access control and permissions. Tokens are created by the manufacturer, or current owner, and this token is combined with the RFID identification of the device. This mechanism ensures the change of permissions by the device itself if it is assigned a new owner, or it is going to be operated in a different department of the same company, thus reducing the overhead of the new owner. These tokens can be changed by the owners, provided that old token is provided, so as to supersede the permissions and access control of the previous one.

6.5 SECURITY ATTACKS IN IoT

The various attacks on the data, device and the channel in the IoT are discussed below.

6.5.1 Malware

Malware presented in Figure 6.20 is malicious software, a kind of program or file that is harmful to a computer user. These malicious programs once entered into user's device, then it can perform a variety of different attacks such as stealing, encrypting or deleting sensitive data, altering or hijacking core computing functions and monitoring users' computer activity without their permission.

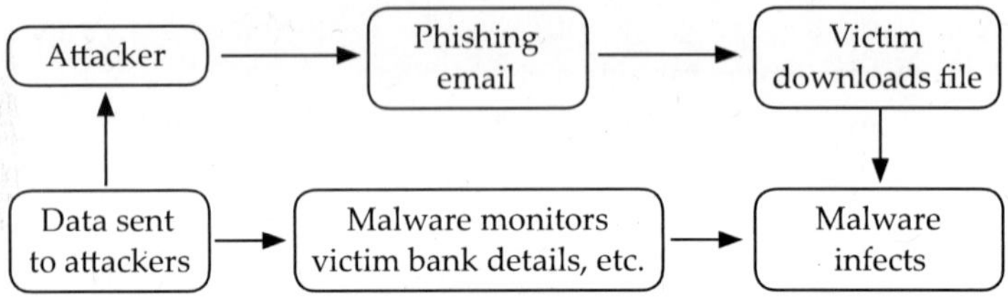

Fig. 6.20 Malware

Working of malware

Malware creators use a variety of physical and virtual methods to spread malware that infects devices and networks. For example, malicious programs can be delivered to a system with a USB drive or can spread over the Internet through drive-by downloads, which automatically download malicious programs to systems without the user's approval or knowledge. Phishing attacks are another common type of malware delivery where emails disguised as legitimate messages contain malicious links or attachments that can deliver the malware executable to unsuspecting users.

Common types of malware

- A virus is the most common type of malware which can execute itself and spread by infecting other programs or files.
- Spyware is made to collect information and data on the device user and observe their activity without their knowledge.
- Ransomware is designed to infect a user's system and encrypt the data. Cyber criminals then demand a ransom payment from the victim in exchange for decrypting the system's data. So the user can't read their own data.
- A backdoor virus or remote access Trojan (RAT) secretly creates a backdoor into an infected system that allows threat actors to remotely access it without alerting the user or the system's security programs.
- Adware is used to track a user's browser and download history with the intent to display pop-up or banner advertisements that lure the user into making a purchase. For example, an advertiser might use cookies to track the web pages a user visits to better target advertising.

- Key loggers, also called system monitors, are used to see nearly everything a user does on their computer. This includes emails, opened web pages, programs and keystrokes.

Examples

- **ILOVEYOU**, a worm that spread like wildfire in 2000 and did more than $15 billion devices in damage.
- **SQL Slammer**, which ground Internet traffic to a halt within minutes of its first rapid spread in 2003.
- **Conficker** is a worm that exploited unpatched flaws in Windows and leveraged a variety of attack vectors from injecting malicious code to phishing email, to ultimately crack passwords and hijack Windows devices into a botnet.
- **Zeus**, a late 00s key logger Trojan that targeted banking information.
- **Stuxnet**, an extremely sophisticated worm that infected computers worldwide but only did real damage in one place: the Iranian nuclear facility at Natanz, where it destroyed uranium enriching centrifuges, the mission it was built for by US and Israeli intelligence agencies

6.5.2 Ransomware

Ransom malware, or ransomware, is a type of malware that prevents users from accessing their system or personal files and demands ransom payment in order to regain access. The earliest variants of ransomware were developed in the late 1980s, and payment was to be sent via snail mail. Today, ransomware authors order that payment be sent via crypto currency or credit card. Figure 6.21 shows the ransomware model.

Fig. 6. 21 Flow of ransomware

Types of ransomware

- **Crypto malware:** This form of ransomware can cause a lot of damage because it encrypts things like your files, folders, and hard drives.
- **Lockers:** These are known for infecting your operating system to completely lock you out of your computer or devices, making it

impossible to access any of your files or applications. This type of ransomware is most often Android based.

- **Scareware:** Scareware is fake software that acts like an antivirus or a cleaning tool. Scareware often claims to have found issues on your computer, demanding money to resolve the problems. Some types of scareware lock your computer.
- **Doxware:** Commonly referred to as leakware or extortionware, doxware threatens to publish your stolen information online if you don't pay the ransom. It's understandable that some people panic and pay the ransom when their files are hijacked.
- **RaaS (ransomware as a service):** RaaS is a type of malware hosted anonymously by a hacker. These cyber criminals handle everything from distributing the ransomware and collecting payments to managing decryptors software that restores data access in exchange for their cut of the ransom.
- **Mac ransomware:** Mac operating systems were infiltrated by their first ransomware in 2016. Known as KeRanger, this malicious software infected Apple user systems through an app called "transmission", which was able to encrypt its victims' files after being launched.
- **Ransomware on mobile devices:** Ransomware began infiltrating mobile devices on a larger scale in 2014. Mobile ransomware is often delivered via a malicious app, which leaves a message on your device that says it has been locked due to illegal activity.

Examples

- **WinLock in 2007:** Rather than encrypting files, this form of ransomware locked its victims out of their desktops and then displayed pornographic images on their screens. In order to remove the images, victims had to pay a ransom with a paid SMS.
- **Reveton in 2012:** This so-called law enforcement ransomware locked its victims out of their desktops while showing what appeared to be a page from an enforcement agency such as the FBI. This fake page accused victims of committing crimes and told them to pay a fine with a prepaid card.
- **CryptoLocker in 2013:** Ransomware tactics continued to progress, especially by 2013 with this military-grade encryption that used key storage on a remote server. These attacks infiltrated over 2,50,000 systems and reaped $3 million before being taken offline.
- **Locky in 2016:** The Locky ransomware used social engineering to deliver itself via email. When it was first released, potential victims were enticed to click on an attached Microsoft Word document, thinking the

attachment was an invoice that needed to be paid. But the attachment contained malicious macros. More recent Locky ransomware has evolved into the use of JavaScript files, which are smaller files that can more easily evade anti-malware products.

- **WannaCry in 2017:** These more recent attacks are examples of encrypting ransomware, which was able to spread anonymously between computers and disrupt businesses worldwide.
- **Sodinokibi in 2019.** The cyber criminals who created this ransomware used managed service providers (MSPs) like dental offices to infiltrate victims on a larger scale.

6.5.3 MITM Attack

A man in the middle (MITM) attack is a general term when a perpetrator positions himself in a conversation between a user and an application is shown in Figure 6.22, either to eavesdrop or to impersonate one of the parties, making it appear as if a normal exchange of information is underway. The goal of an attack is to steal personal information, such as login credentials, account details and credit card numbers. Targets are typically the users of financial applications, SaaS businesses, e-commerce sites and other websites where login is required. Information obtained during an attack could be used for many purposes, including identity theft, unapproved fund transfers or an illicit password change.

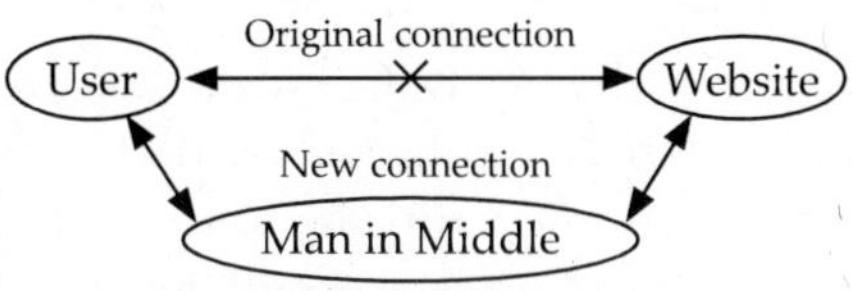

Fig. 6.22 MITM attack

MITM attack progression

Successful MITM execution has two distinct phases: interception and decryption.

Interception

The first step intercepts user traffic through the attacker's network before it reaches its intended destination. The most common (and simplest) way of doing this is a passive attack in which an attacker makes free, malicious Wi-Fi hotspots available to the public. Typically named in a way that corresponds to their location, they aren't password protected. Once a victim connects to such a hotspot, the attacker gains full visibility to any online data exchange.

Attackers wishing to take a more active approach to interception may launch one of the following attacks.

- IP spoofing involves an attacker disguising himself as an application by altering packet headers in an IP address. As a result, users attempting to access a URL connected to the application are sent to the attacker's website.
- ARP spoofing is the process of linking an attacker's MAC address with the IP address of a legitimate user on a local area network using fake ARP messages. As a result, data sent by the user to the host IP address is instead transmitted to the attacker.
- DNS spoofing also known as DNS cache poisoning, involves infiltrating a DNS server and altering a website's address record (Figure 6.23). As a result, users attempting to access the site are sent by the altered DNS record to the attacker's site.

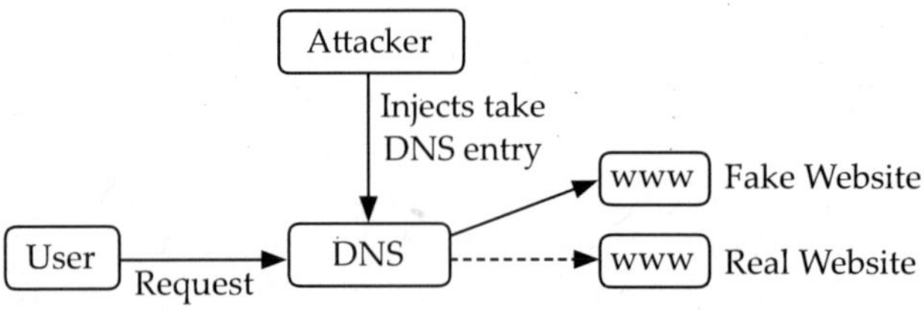

Fig. 6.23 DNS spoofing

Decryption

After interception, any two-way SSL traffic needs to be decrypted without alerting the user or application. A number of methods exist to achieve this.

- HTTPs spoofing sends a phony certificate to the victim's browser once the initial connection request to a secure site is made. It holds a digital thumbprint associated with the compromised application, which the browser verifies according to an existing list of trusted sites. The attacker is then able to access any data entered by the victim before it's passed to the application.
- SSL BEAST (Browser Exploits Against SSL/TLS) targets a TLS version 1.0 vulnerability in SSL. Here, the victim's computer is infected with malicious JavaScript that intercepts encrypted cookies sent by a web application. Then the app's cipher block chaining (CBC) is compromised so as to decrypt its cookies and authentication tokens.
- SSL hijacking occurs when an attacker passes forged authentication keys to both the user and application during a TCP handshake. This sets up what appears to be a secure connection when, in fact, the man in the middle controls the entire session.

- SSL stripping downgrades a HTTPS connection to HTTP by intercepting the TLS authentication sent from the application to the user. The attacker sends an unencrypted version of the application's site to the user while maintaining the secured session with the application. Meanwhile, the user's entire session is visible to the attacker.

6.5.4 Mirai Attack

Mirai is a self-propagating botnet virus. The source code for Mirai was made publicly available by the author after a successful and well publicized attack on the Krebbs website. Since then the source code has been built and used by many others to launch attacks on Internet infrastructure. The Mirai botnet code infects poorly protected Internet devices by using telnet to find those that are still using their factory default username and password. The effectiveness of Mirai is due to its ability to infect tens of thousands of these insecure devices and coordinate them to mount a DDOS attack against a chosen victim. The work flow of Mirai attack is shown in Figure 6.24.

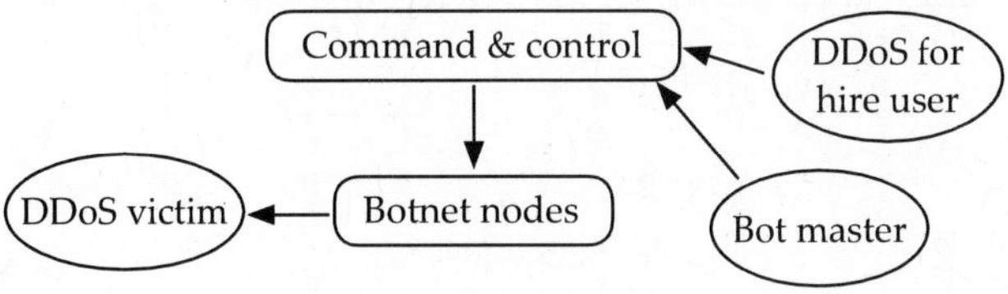

Fig. 6.24 Mirai

(https://www.imperva.com/blog/how-to-identify-a-mirai-style-ddos-attack/)

Working of Mirai

There are two main components of Mirai, the virus itself and the command and control center (CnC). The virus contains the attack vectors, Mirai has ten vectors that it can launch, and a scanner process that actively seeks other devices to compromise. The CnC is a separate image that controls the compromised devices (BOT) sending them instructions to launch one of the attacks against one or more victims.The scanner process runs continuously on each BOT using the telnet protocol (on TCP port 23 or 2323) to try and login to IP addresses at random. The login tries up to 60 different factory default username and password pairs when login succeeds the identity of the new BOT and its credentials are sent back to the CnC. The CnC supports a simple command line interface that allows the attacker to specify an attack vector, a victim(s) IP address and an attack duration. The CnC also waits for its existing BOTs to return newly discovered device addresses and credentials which it uses to copy over the virus code and, in turn, creates new BOTs.

What is an IoT botnet?

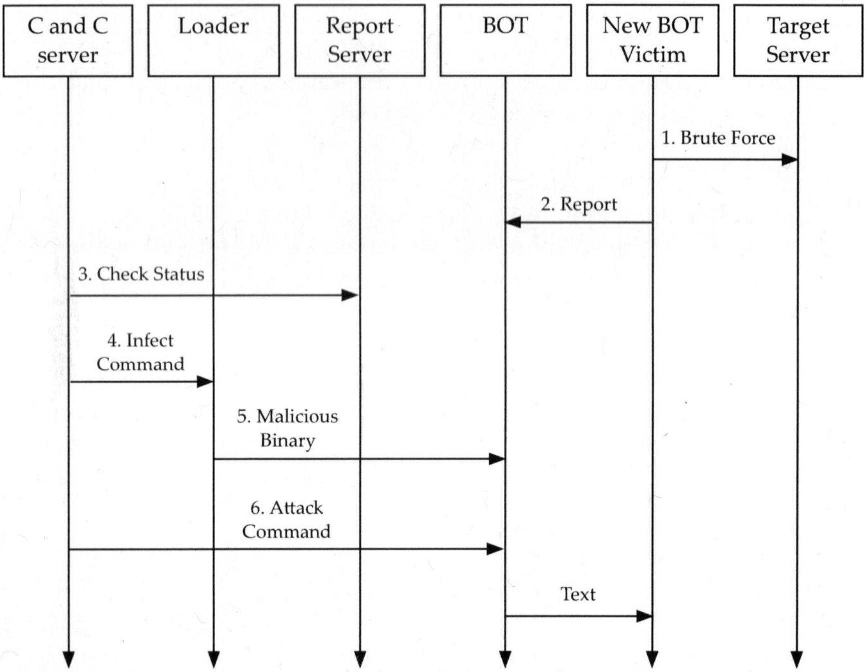

Fig. 6.25 IoT botnet

Over the years, PC makers worried about building security into their computers. But another tempting target is out there for botnet builders. Figure 6.25 gives the detailed working process of IoT Botnet like how the Botnet attack arises and how the commands have been followed. Internet of things (IoT) devices, a blanket term for various gadgets that most people don't think of as computers, but that still have processing power and an Internet connection. These devices, ranging from home routers to security cameras to baby monitors, often include an embedded, stripped down Linux system. They also often have no built-in ability to be patched remotely and are in physically remote or inaccessible locations.

By 2017, there were 8.4 billion of these "things" out there on the Internet, ripe for plucking. Mirai took advantage of these insecure IoT devices in a simple but clever way. Rather than attempting to use complex wizardry to track down IoT gadgets, it scanned big blocks of the Internet for open Telnet ports, then attempted to log in using 61 username/password combos that are frequently used as the default for these devices and never changed. In this way, it was able to amass an army of compromised closed-circuit TV cameras and routers, ready to do its bidding.

6.6 SECURITY VULNERABILITIES IN IoT

The severity of impact depends greatly on the individual system, the collected data and its information. For example, an attack disabling the brakes of a connected car, or on a connected health device, such as an insulin pump hacked to administer too much medication to a patient, can be life threatening. Likewise, an attack on a refrigeration system housing medicine that is monitored by an IoT system can ruin the viability of a medicine if temperatures fluctuate. Similarly, an attack on critical infrastructure like an oil well, energy grid or water supply can be disastrous means create great damage. Figure 6.26 shows the various reasons for security vulnerabilities in IoT. It depends on operational view cost and lack of updating on software.

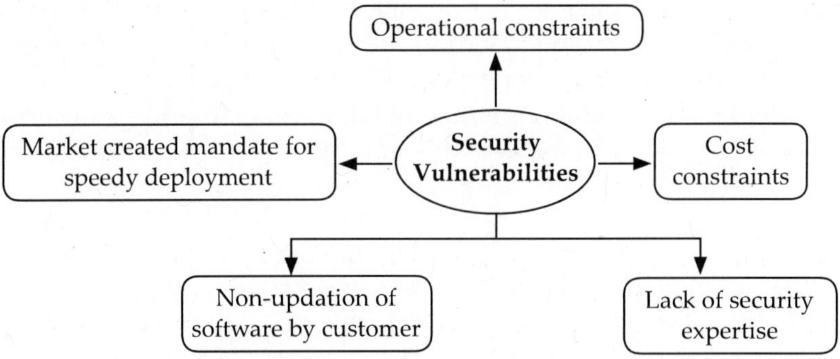

Fig. 6.26 Reason for security vulnerabilities in IoT

Other attacks, however, cannot be underestimated. For example, an attack against smart door locks could potentially allow a burglar to enter a smart home. In other scenarios such as the 2013 Target hack or other security breaches, an attacker could pass malware through a connected system an HVAC system in Target's case to scrape personally identifiable information, wreaking havoc for those affected.

6.6.1 IoT Security Breaches and Hacks

The IoT breaches and hacks occur when unauthorized user enters and leaves from different network. Security experts have warned of the potential risk of large number of unsecured devices connected to the Internet since the IoT concept first originated in the late 1990s. A number of attacks subsequently have made headlines, from refrigerators and TVs being used to send spam to hackers infiltrating baby monitors and talking to children. It is important to note that many of the IoT hacks don't target the devices themselves, but rather use IoT devices as an entry point into the larger network.

For example in 2010, researchers revealed that the Stuxnet virus was used to physically damage Iranian centrifuges, with attacks starting in 2006 but the primary attack occurring in 2009. Often considered one of the earliest examples of an IoT attack, Stuxnet targets supervisory control and data acquisition (SCADA) systems in industrial control systems (ICS), using malware to infect instructions sent by programmable logic controllers (PLCs). Attacks on industrial networks have only continued with malware such as Crash Override/Industroyer, Triton and VPN Filter targeting vulnerable OT and industrial IoT systems.

In December 2013, a researcher at enterprise security firm Proofpoint Inc. discovered the first IoT botnet. According to the researcher, more than 25% of the botnet was made up of devices other than computers, including smart TVs, baby monitors and household appliances. In 2015, security researchers Charlie Miller and Chris Valasek executed a wireless hack on a jeep, changing the radio station on the car's media center, turning its windshield wipers and air conditioner on, and stopping the accelerator from working. They said they could also kill the engine, engage the brakes and disable the brakes altogether. Miller and Valasek were able to infiltrate the car's network through Chrysler's in vehicle connectivity system, Uconnect.

Mirai, one of the largest IoT botnets to date, first attacked journalist Brian Krebs website and French web host OVH in September 2016. OVH expanses On Vous Héberge in French Language of "We Host You" meaning. The attacks clocked in at 630 gigabits per second (Gbps) and 1.1 terabits per second (Tbps), respectively. The following domain name system (DNS) service provider Dyn's network was targeted, making a number of websites, including Amazon, Netflix, Twitter and The New York Times, unavailable for hours. The attacks infiltrated the network through consumer IoT devices, including IP cameras and routers. A number of Mirai variants have since emerged, including Hajime, Hide N Seek, Masuta, PureMasuta, Wicked botnet and Okiru, among others. In January 2017 notice, the Food and Drug Administration (FDA) warned that the embedded systems in radio frequency enabled St. Jude Medical implantable cardiac devices, including pacemakers, defibrillators and resynchronization devices, could be vulnerable to security intrusions and attacks.

6.6.2 IoT Security Standards

The policies and standards should ensure that data will be managed, protected, and transmitted in an efficient way. Service level agreements (SLAs) must be clearly identified in every service involved. Current policies that are used in computer and networks security may not be applicable for IoT due to its heterogeneous and dynamic nature. The enforcement of such policies will introduce trust by human users in the IoT paradigm which will eventually result in its growth and scalability.

Universal policies and standards to control the design and the implementation of algorithms in IoT make it difficult to control the security. It is important for IoT architecture to have a federated architecture that has an internal autonomy or centralized unit to overcome the heterogeneity of various devices, software and protocols. Another such attempt was made by the authors to propose a framework called Secure Mediation GateWay (SMGW) for critical infrastructures. This approach is an abstraction of IoT as it is relevant for any kind of distributed infrastructures that are completely different in their nature and operation. SMGW can discover all the relevant distributed information from different nodes, and can overcome the heterogeneity of heterogeneous nodes whether it is a telecommunication, electrical, water distribution node, and can exchange all the messages and information over the untrusted network of Internet. It is not enough to have policies and standards to ensure security but mechanisms to enforce such policies are also needed. The research by Neisse et al., addresses this issue by integrating a security toolkit named SecKit with the MQ Telemetry Transport protocol. The current policies may not be efficient in IoT because of its dynamic nature. The proposed policy mechanism can have good impact in ensuring the security of the IoT. However, it introduced additional delay in the process.

Many IoT security frameworks exist, but there is no single industry accepted the standard to date. However, simply adopting an IoT security framework can help. They provide tools and checklists to help companies creating and deploying IoT devices. Such frameworks have been released by GSM Association, the IoT Security Foundation, the Industrial Internet Consortium and others. In September 2015, the Federal Bureau of Investigation (FBI) released a public service announcement, FBI Alert Number I-091015-PSA, which warned about the potential vulnerabilities of IoT devices and offered consumer protection and defense recommendations.

In August 2017, US Congress introduced the IoT Cyber Security Improvement Act, which would require any IoT device sold to the US government not to use default passwords, not have known vulnerabilities and offer a mechanism to patch the devices. While aimed at those manufacturers creating devices being sold to the government, it set a baseline for security measures all manufacturers should adopt. Also in August 2017, the Developing Innovation and Growing the Internet of Things (DIGIT) Act passed by the Senate. This bill would require the

Department of Commerce to convene a working group and create a report on IoT, including security and privacy.

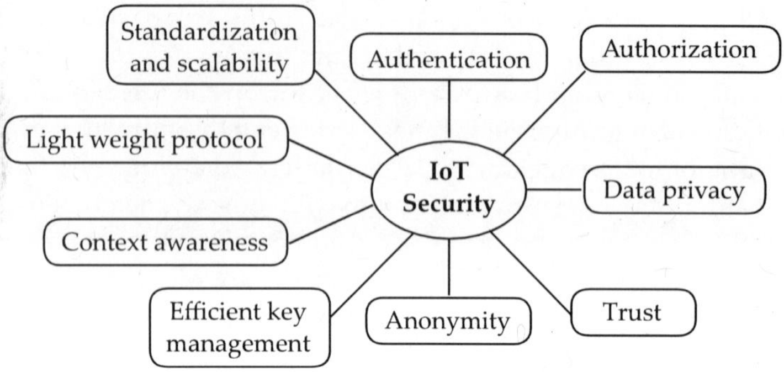

Fig. 6.27 IoT security

While not IoT-specific, the General Data Protection Regulation (GDPR), released in May 2018, unifies data privacy laws across the European Union. These protections extend to IoT devices and their networks and IoT device makers should take them into account. In June 2018, Congress introduced the State of Modern Application, Research and Trends of IoT Act, or SMART IoT Act, to propose the Department of Commerce to conduct a study of the IoT industry and provide recommendations for the secure growth of IoT devices. In September 2018, California state legislature approved SB-327 Information privacy: connected devices, a law that introduced security requirements for IoT devices sold in the country. Figure 6.27 shows the IoT security which is developed to provide authentication, authorization, data privacy, trust, standardization and scalability, context awareness and efficient key management.

The Union Cabinet chaired by Prime Minister Shri Narendra Modi approved the launching of National Mission on Interdisciplinary Cyber-Physical Systems (NM-ICPS) to be implemented by Department of Science and Technology at a total outlay of Rs. 3660 crore for a period of five years in 2019.

Cyber physical systems (CPS) is a new class of engineered systems that integrates computation and physical processes in a dynamic environment. CPS encompasses technology areas of Cybernetics, Mechatronics, Design and Embedded systems, Internet of Things (IoT), Big Data, Artificial Intelligence (AI) and many more. The CPS systems are intelligent, autonomous and efficient and are expected to drive innovation in sectors as diverse as agriculture, water, energy, transportation, infrastructure, security, health and manufacturing. Thus, it is heralded as the next paradigm shift in technology that can exponentially spur growth and development. The mission will be implemented through a network of 15

Technology Innovation Hubs (TIHs), 6 Sectoral Application Hubs (SAHs) and 4 Technology Translation Research Parks (TTRPs). Each hub and technology park will follow a technology lifecycle approach, addressing all stages.

6.6.3 Application Level Security Challenges

Depending on the topology and devices in your home that connects to the Internet, things will get worse. Let's take the smart assistants like Alexa, they have speakers that connect to the Internet. They need to be affordable and power efficient so extra SW tasks and extra HW tasks for security are not the first priority. Especially when microphones or cameras are involved, it will be interesting to see how a growing amount of Internet pirates will make use of this to extort people.

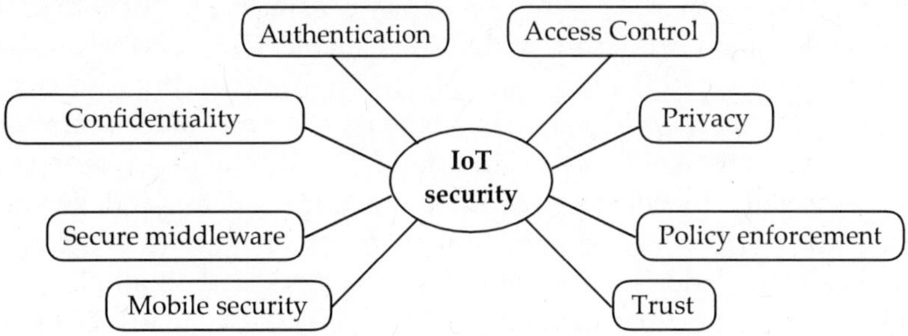

Fig. 6.28 IoT security challenges

A number of challenges prevent the securing of IoT devices as shown in Figure 6.28 and ensuring end-to-end security in an IoT environment. Because the idea of networking appliances and other objects is relatively new, security has not always been considered top priority during a product's design phase. Additionally, IoT is a booming field, so nascent market, many product designers and manufacturers are more interested in getting their products to market quickly, rather than taking the necessary steps to build security in from the start.

Major challenges cited with IoT security is the use of hardcoded or default passwords, which can lead to security breaches. Even if passwords are changed, they are often not strong enough to prevent infiltration.

Many devices do not or cannot offer advanced security features. For example, sensors that monitor humidity or temperature cannot handle advanced encryption or other security measures. Also as many IoT devices are "set it and forget it" placed in the field or on a machine and left until end of life, they hardly ever receive security updates or patches. From a manufacturer's viewpoint, building security from the start can be costly, slow down development and cause the device not to function as it should.

6.6.4 Legacy

Connecting legacy assets not inherently designed for IoT connectivity are another security challenge. Replacing legacy infrastructure with connected technology is cost prohibitive, so many assets will be retrofitted with smart sensors. However, as legacy assets that likely have not been updated or ever had security against modern threats, the attack surface is expanded. In terms of updates, many systems only include support for a set timeframe. For legacy and new assets, security can lapse if extra support is not added. Many IoT devices stay in the network for many years, for which adding security can be challenging.

6.6.5 Industry

Large companies and industry organizations may have their own specific standards, while certain segments, such as industrial IoT, have proprietary, incompatible standards from industry leaders. The variety of these standards makes it difficult to not only secure systems, but also ensure interoperability between them. The convergence of IT and operational technology (OT) networks has created a number of challenges for security teams, especially those tasked with protecting systems and ensuring end-to-end security in areas outside their realm of expertise. A learning curve is involved, and IT teams with the proper skill sets should be put in charge of IoT security. The major developments witnessed in IoT are mainly on small scale, i.e., within companies, some industries, etc. To scale the IoT framework from one company to a group of different companies and systems, various security concerns need to be overcome. The IoT has great potential to transform the way we live today. But, the foremost concern in realization of completely smart frameworks is security. If the security concerns like privacy, confidentiality, authentication, access control, end-to-end security, trust management, global policies and standards are addressed completely, then we can witness the transformation of everything by IoT in the near future. There is need for new identification, wireless, software, and hardware technologies to resolve the currently open research challenges in IoT like the standards for heterogeneous devices, implementation of key management and identity establishment systems, and trust management hubs.

6.7 ONGOING RESEARCH IN IoT TECHNOLOGY

IoT security is determined by many factors and security principles discussed earlier, and the challenges that are faced by IoT security has been the focus of many researchers. In the survey paper presented by Roman (et al). a detailed introduction about the IoT and security issues along with the need to have IoT standards are addressed. IoT security in terms of the main principles of security like confidentiality, integrity, and availability are addressed only. The authors suggested two-step authorization using biometrics which is not applicable in case-of-machine-to-machine communication. The suggested measures are not detailed and do not address the specific nature of IoT with low power heterogeneous devices and huge network traffic. Wireless Internet Service Provider (WISPr) roaming and RADIUS are existing solutions to provide authentication and authorization in IoT by means of Wi-Fi over the Internet. Today, many smart devices support IPV6 communications, but the existing deployments in IoT might not support it, and thus require ad hoc gateways and middleware.

Future Directions

IoT has seen rapid development in recent years in the areas of telemedicine platforms, intelligent transportation systems, logistics monitoring, and pollution monitoring systems, agriculture monitoring, industry automation, etc. However, the security challenges related to the IoT must be dealt with to achieve its growth and maturation. The future directions for research is in order to make the Io'T paradigm more secure with the standard architecture, additional layer, new standard and technique, advances in communication, speed, bandwidth, etc.

Architecture Standards

IoT currently employs different devices, services, and protocols to achieve a common goal. However, to integrate a network with different IoT frameworks to achieve a bigger framework, for example, to form a smart town by the integration of many smart homes, there needs to be a set of standards that should be followed from the micro to macro levels of IoT realization. The present-day requirement of IoT is to have well defined architecture standards comprising of data models, interfaces, and protocols which can support a wide range of humans, devices, languages, and operating systems.

To realize the implementation of IoT IPV4 will definitely fall short in accommodating the huge numbers of IP-identifiable objects. That is the reason why people are now heading to IPV6, which is able to support devices. However, such number will create huge amount of traffic, which can lead to more delay and thus more bandwidth is needed. The expectation of the new generation of communication (5G) is to provide speed between 10-800 Gbps, comparing this number with the current technology (4G) with speed of 2–1000 Mbps, 5G should be able to handle the traffic produced by IoT devices. 5G technology is also expected to accommodate both IPV4 and IPV6 by having IPV4/IPV6 framework translation. The implementation of 5G will be defined by many current and developing technologies such as: Heterogeneous Networks (HetNets), Software Defined Networks (SDNs), Massive MIMO, and Multiple Radio Access, etc. However, all these technologies come with their own security challenges. For example, HetNets will have frequent handover which directly affects the authentication process in the network, especially with the small latency requirement of 5G. Also, Cloud computing and SDNs will increase the numbers of DDoS attacks due to the On-Demand Self Service characteristic of Cloud computing. Although authentication and security of SDN are addressed by having a decentralized control of authentication using user-dependent security context, the security of 5G and all the emerging technologies involved in 5G must be extensively addressed, in order to ensure IoT security.

Current state of research in IoT is mainly focused on authentication and access control protocols, but with the rapid advancement of technology it is essential to incorporate new networking protocols like IPV6 and 5G to achieve the dynamic smashup of IoT topology. Because of that, the 5G will support in different speed with different bandwidth, different Internet standard which are adaptable to access the multiple networks. The first phase of NM-ICPS will focus on establishing six Technology Innovation Hubs (TIHs) in the following domain areas:

- Artificial Intelligence and Machine Learning
- Technologies for Internet of Things and Everything (IOT & IOE), Sensors, Activators and Control
- Databanks and Data Services, Data Analytics
- Advanced Communication Systems
- Robotics and Autonomous Systems
- Cyber Security and Cyber Security for Physical Infrastructure

The first phase of NM-ICPS will be implemented by Science and Engineering Research Board (SERB). Proposals are invited from reputed Academic and R&D institutions, with an impressive track record of scientific excellence, depth of experience in the proposed field and a strategic vision, to establish and sustain dynamic technology innovation hubs in cyber physical systems.

6.8 SUMMARY

In this chapter different demands of security measures and challenges were discussed in device level and network level. The various possibilities of attacks and vulnerability for data in all IoT layers were illustrated. And finally ongoing research of security for IoT was discussed.

REVIEW QUESTIONS

1. What is the need of security in IoT?
2. What is meant by security mechanism?
3. What is CIA?
4. What are the security measures used in IoT?
5. Discuss the three levels of security with examples.
6. Describe the security challenges faced in network layer.
7. What types of attack are possible in IoT?
8. Explain MITM attack in detail.
9. Define ransomware.
10. What is the function of firewall?
11. List the security vulnerabilities present in IoT.
12. What standards are available for IoT security?

PART-II
Developing Internet of Things

7

IoT IMPLEMENTATION TOOLS

7.1 INTRODUCTION

This chapter focuses on practical knowledge for implementing IoT tools, hardware facilities with specifications for Raspberry Pi, Arduino with python code. Interfacing of IoT with cloud and its API concepts are briefed.

7.2 PYTHON

Guido Van Rossum who was the creator of Python faced some shortcomings of ABC language (namely, exception handling) while working in the Amoeba distributed operating system at Centrum Wiskunde and Informatica (CWI) situated in the Netherlands decided to create some user-friendly language and designed the Python coding in 1980. Then, Rossum started to work on developing a new language that integrated all good features of ABC language and new desired features, such as extensibility and exception handling. As a successor of other programming languages the first version of Python programming language was released in 1992 by CWI in the Netherlands.

Python is an easy to learn programming language widely used to develop scalable web applications. It provides excellent library support and has a large developer community. This language provides a great and easy starting point for beginners. As a description of its easiness, a sample code to print "Hello World!" is given: print "Hello World!". It is syntax to display Hello word, see how simple and user friendly it is! Certain web applications like YouTube, Instagram, Pinterest, and Survey Monkey have been built in Python.

Python 1.0 version was released in 1994 with new features such as lamda map, etc. Python 2.0 version was released in 2000 with several improvements. Python 3.0 version was released in 2008 to help the programmers to develop projects easily. Recently, in 2020 python version 3.9 was released to help the programmers with advanced techniques like any web based applications.

7.2.1 Features of Python

1. **Interpreter:** Python is an interpreted language. Here the source code is translated into byte code unlike machine code conversion in C/C^{++}. The byte code is a low level set of instructions that can be executed by an interpreter on a virtual machine.
2. **Platform independent:** Python is a platform independent programming language. It means Python can easily run on a variety of platforms like Windows, Linux and UNIX. No need to write a separate code for each operating system, single code can run on all the multiple platforms.
3. **Open source:** Another useful feature of Python is that it is open source. Using the Internet, anyone can easily download and use Python.
4. **High level language:** This is one key feature of Python. This means that there is no necessity for the programmers to remember system architecture and manage memory.
5. **Library support:** Large set of Python libraries are available to programmers that help them to easily write the code. A large set of libraries means that there is no need for programmers to write a code for every single thing.
6. **Object oriented concept:** Python supports object oriented concept. This feature provides a lot of flexibility for the developers to easily develop various projects.
7. **Supports exception handling:** An exception is an event that occurs during execution of a program, exception can disrupt the normal flow of the program. Python supports exception handling which means the programmer can write less error-prone code and can test various scenarios that can cause an exception later on.
8. **Automatic memory management:** Python supports automatic memory management which means the memory is cleared and freed automatically. The programmer need not have to bother clearing the memory.
9. **Dynamically typed language:** Python is a dynamically typed language. That means the data type (for example, int, double, long, etc.) for a variable is decided at runtime and not in advance because of this feature no need to specify the type of variable.
10. **Extensible feature:** Python is an extensible language. We can write a Python code into C or C^{++} language and also compile that code in C/C^{++} language.

7.2.2 Various Python Implementations

Python is an interpreted, high level and general-purpose programming language. Its object oriented approach helps programmers to write clear, logical code for small and large scale projects. It can support multiple programming paradigms included structured (particularly, procedural), objected oriented and functional programming. It is often described as a "batteries included" language due to its comprehensive standard library. It has a great backend for web development, data analysis, artificial intelligence, and scientific computing.

Python has multiple implementations includes the following:

- **CPython**, which is the default interpreter for Python which is written in **C programming language** and developed by Python Software Foundation.
- **Jython** is an implementation of Python interpreter scripted in **Java programming language** for Java Virtual Machine (JVM).
- **IronPython** is an implementation of the Python programming language targeting the .NET framework and Mono wirtten in C#.
- **PyPy** is an implementation of Python which does not use an interpreter. It is implemented using just-in-time compiler. It is written in RPython and translated into C.
- **Stackless Python** is CPython with co routines.
- **MicroPython** uses Python 3 implementation for microcontroller platforms.

Python is also capable of interacting with other languages by using certain modules. Most of these modules work on community development model and are open source.

Some of the Python Integrated Development Environment (IDE) include:

- **Atom**, an open source cross platform IDE with auto complete, help and more Python features under package extensions.
- **Easy Eclipse**, an open source IDE for Python and other languages.
- **Eclipse** with the Pydev plug in. Eclipse supports many other languages as well.
- **NetBeans**, is written in Java and runs everywhere where a JVM is installed.
- **PyScripter**, open source software Python IDE for Windows.
- **Spyder**, IDE for scientific programming.

7.2.3 Execution Process of Python

When the Python software is installed on a machine, minimally, it has:

- An interpreter
- A support library.

What is an Interpreter?

Interpreter is a software that is used to run the Python scripts. An interpreter function is given in Figure 7.1 which executes the source code line by line. C Python is the default interpreter for Python which is written in C programming language.

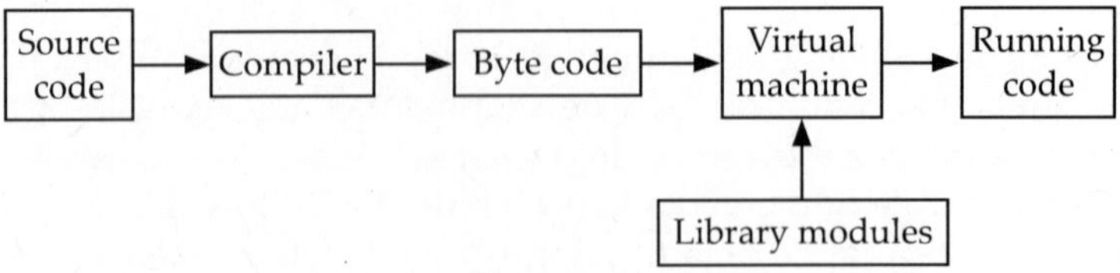

Fig. 7.1 Interpreter

Python Interpreter is made up of two parts, namely, compiler and virtual machine. The Python interpreter performs the following two steps to execute a Python program.

STEP1: Compilation

Compilation is the process of translation. The interpreter reads a python code or instruction and verifies whether the instruction is well formatted, i.e., it checks the syntax of each line. If it encounters any error, it immediately halts the translation and shows an error message. If there is no error, i.e., if the python instruction or code is well formatted then the compiler translates the source code into its equivalent form in intermediate language called "Byte code". Byte code is a lower level, platform independent, efficient and intermediate representation of the source code. The byte code makes Python platform independent. The process of compilation in C Python interpreter's compiler can be divided into 4 main parts. They are discussed one by one

- **Parse source code into a parse tree:** Based on grammar rules of Python programming language, the source code is converted to a parse tree. Every node of the parse tree contains a part of code. Consider the given simple arithmetic expression [14 + (2 × 3) – (6 / 2)] and its parse tree is drawn in Figure 7.2:

14 + (2 * 3) – (6 / 2)

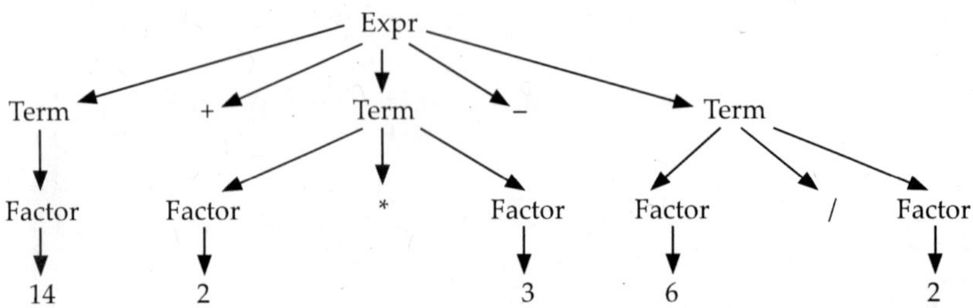

Fig. 7.2 Parse tree for example

- **Transform parse tree into an abstract syntax tree (AST):** The abstract syntax tree (AST) is a high level representation of the program structure. Each node of the tree denotes a construct in occurrence in the source code. The syntax is "abstract" in not representing every detail appearing in the real syntax. Consider the AST as shown in Figure 7.3 for the parse tree example.

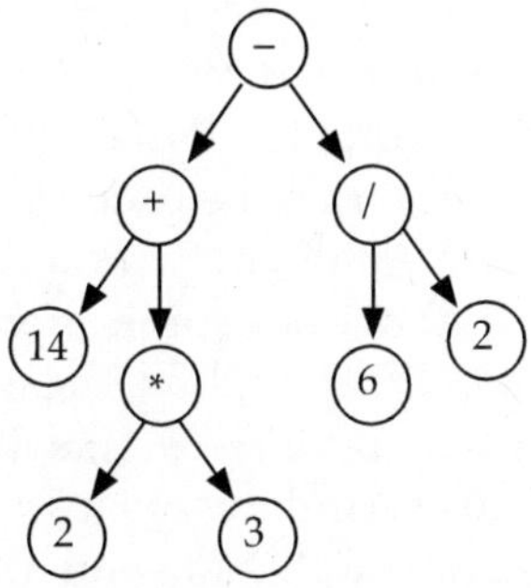

Fig. 7.3 AST structure

- **Transform AST into a control flow graph (CFG):** A control flow graph (CFG) is a directed graph that models the flow of a program using basic blocks. Each block contains the byte code representation of program code inside it. Byte code is directly generated from the basic blocks by doing a post-order depth first search on the CFG following the edges. CFGs are usually one step away from final code output.

STEP 2: Python Virtual Machine

The source code gets converted to byte code and then it is fed to PVM (python virtual machine). It's a big loop that iterates through the byte code instructions, one by one, to carry out their operations. The PVM is the runtime engine of Python. It's always present as part of the Python system and is the component that truly runs the scripts. Technically, it's the last step of what is called the Python interpreter. If an error occurs during this execution then the execution is halted with an error message.

7.2.4 Applications of Python

Nowadays python based applications are developed tremendously. A few of the applications are listed.

1. **Web development:** Web framework like Django and Flask are based on Python. They are used to write server side code which helps to manage database, write backend programming logic, mapping URLs, etc.
2. **Machine learning:** There are many machine learning applications written in Python. Machine learning is a way to write logic so that

a machine can learn and solve a particular problem on its own. For example, products recommendation in websites like Amazon, Flipkart, eBay, etc., is a machine learning algorithm that recognises user's interest. Face recognition and voice recognition in phone is another example of machine learning.

3. **Data analysis:** Data analysis and data visualisation in the form of charts can also be developed using Python.
4. **Scripting:** Scripting is writing small programs to automate simple tasks such as sending automated response emails, etc. Such type of applications can also be written in Python programming language.
5. **Game development:** Python can be used to develop games.
6. **Embedded systems and robotics:** Python can be used to develop codes for embedded and robotics applications.
7. **Desktop applications:** With Python, desktop applications like GUI can be developed using libraries like TKinter or QT. PyQT is available for Unix/Linux, Windows, Mac OS X and Sharp Zaurus. It is combines the best of Python and Qt and it is up to the programmer to decide whether to create a program by coding or using Qt Designer to create visual dialogs.

7.2.5 Python in Embedded Systems

The C/C^{++} programming languages dominate embedded systems program-ming, though it has a number of disadvantages. On the other hand Python has much strength that makes it a great language for embedded systems. While C/C^{++} is slow to write, error prone, and frequently unreadable, Python is known for its writability, error reduction, and readability. Embedded algorithms are getting increasingly complex. Simple control loops and other heavily involved processes are being replaced by neural networks. This has helped Python gain its foothold in embedded system applications. Micro Python is a lean and efficient implementation of the Python 3 programming language with a small subset of the Python standard library and is optimised to run on microcontrollers. Micro Python supports a number of ARM based architectures such as ARM Cortex-M (many STM32 boards, TI CC3200/WiPy, Teensy boards, Nordic nRF series, SAMD21 and SAMD51), ESP8266, ESP32, 16-bit PIC, UNIX, Windows, Zephyr, and JavaScript. Some modern embedded devices have enough memory and a fast enough CPU to run a typical Linux based environment and running C Python on such devices is mostly a matter of compilation (or cross compilation) and tuning. Devices which could be considered 'embedded' by modern standards and which can run tuned versions of CPython include Raspberry Pi, Gumstix, BeagleBone Black.

Advantages of Python

- Creating classes and objects is easy thanks to OOP characteristics.
- Extensive library support.
- Focuses on code readability.
- Has the ability to scale even the most complex applications.
- Ideal for building prototypes and testing ideas faster.
- Open source with an ever growing community support.
- Provides support for a multitude of platforms and systems.
- Easy to learn and use.

Disadvantages of Python

- Not suitable for mobile computing.
- Slower by virtue of being an interpreted programming language.
- The database access layer needs development.
- Threading isn't good because of GIL (global interpreter lock).

7.3 IOT PLATFORMS AND TOOLS

Most frequently used tools to develop IoT applications are listed below.

7.3.1 Arduino

Arduino is an open-source electronics company producing microcontrollers for building digital devices with software support. Arduino boards are able to read inputs, control and producing outputs. There are various versions of Arduino available in market. They are listed with their properties as follows.

Versions of Arduino

(a) **Arduino Uno:** The ATmega328 on the Arduino Uno comes preprogrammed with a boot loader that allows uploading new code to it without the use of an external hardware programmer. It communicates using the original STK500 protocol. The pin diagram in Figure 7.4 and its properties are given in Table 7.1.
The Uno differs from all the preceding boards in that it does not use the FTDI USB-to-serial driver chip. Instead, it features the Atmega16U2 (Atmega8U2 up to version R2) programmed as a USB-to-serial converter.

(b) **Arduino Nano:** The Arduino Nano is designed in a way that allows it to be reset by software running on a connected computer rather than requiring a physical press of the reset button before an upload. One of the hardware flow control lines (DTR) of the FT232RL is connected to the reset line of the ATmega328 via a 100 nano farad

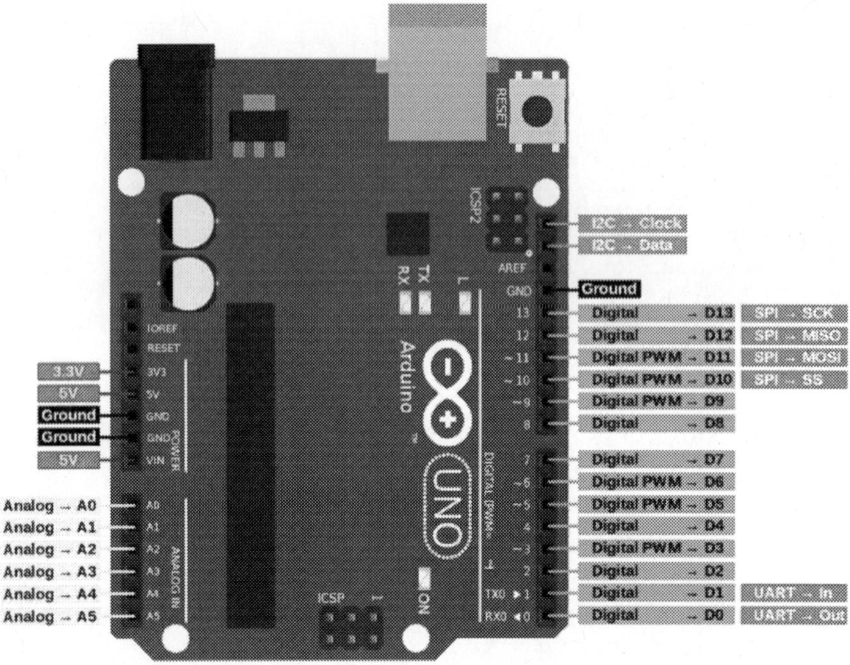

Fig. 7.4 Arduino Uno pin diagram

(https://robu.in/wp-content/uploads/2020/07/1-3.jpg)

Table 7.1. Properties of Arduino Uno

Microcontroller	ATmega328
Operating voltage	5V
Digital I/O pins	14
Analog input pins	6
DC per I/O pin	40 mA
DC for 3.3 V pin	50 mA
Flash memory	32 KB
SRAM	2 KB
EEPROM	1 KB
Clock speed	16 MHz

capacitor. When this line is asserted (taken low), the reset line drops long enough to reset the chip. The pin diagram in Figure 7.5 in its properties are given and Table 7.2.

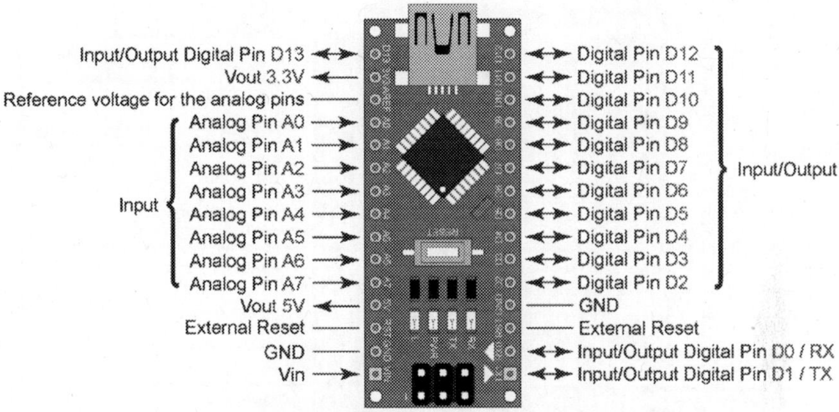

Fig. 7.5 Arduino nano pin diagram

(https://image.makewebeasy.net/makeweb/m_1920x0/EeZNluzYD/Nano_Starter_Kit/ Arduino_nano_pin_1.jpg?v=202012190947)

Table 7.2. Properties of Arduino nano

Microcontroller	ATmega168-20 AU
Operating voltage	5 V
Digital I/O pins	13
Analog input pins	8
DC per I/O pin	40 mA
DC for 3.3 V Pin	50 mA
Flash memory	16 KB
EEPROM	0.5 KB
Clock speed	16 MHz

(c) **Arduino Leonardo:** The Leonardo differs from all the preceding boards in that the ATmega32u4 has built-in USB communication, eliminating the need for a secondary processor. This allows the Leonardo to appear to a connected computer as a mouse and keyboard, in addition to a virtual (CDC) serial / COM port. It also has other implications for the behavior of the board. The pin diagram in Figure 7.6 and its properties are given Table 7.3.

It contains everything needed to support the microcontroller. Its operation for power on is similar to that of Arduino Micro.

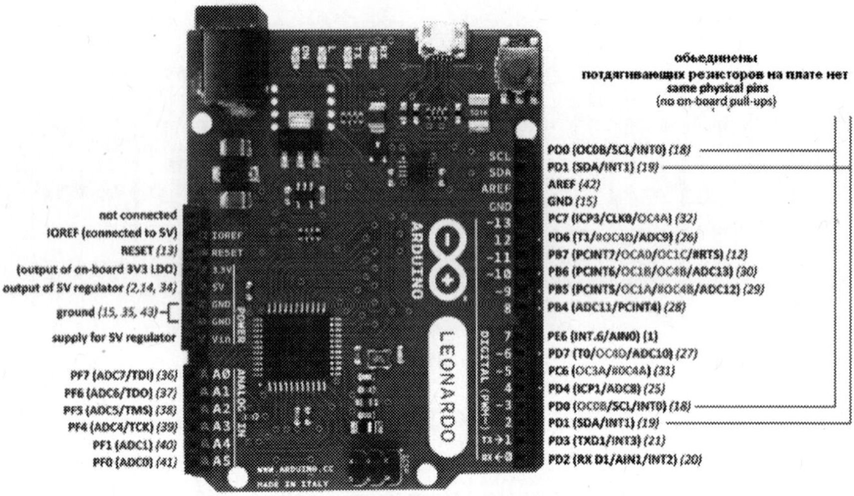

Fig. 7.6 Arduino Leonardo Pin diagram

(https://www.electroschematics.com/wp-content/uploads/2013/01/arduino-leonardo-pinout.jpg?fit=749%2C625)

Table 7.3. Properties of Arduino Leonardo

Microcontroller	ATmega32u4
Operating voltage	5 V
Digital I/O pins	20
PWM channels	7
Analog input pins	12
DC per I/O pin	40 mA
DC for 3.3 V pin	50 mA
Flash memory	32 KB
SRAM	2.5 KB
EEPROM	1 KB
Clock speed	16 MHz

(d) **Arduino Micro:** The Micro is similar to the Arduino Leonardo in that the ATmega32u4 has built-in USB communication, eliminating the need for a secondary processor. This allows the Micro to appear to a connected computer as a mouse and keyboard, in addition to a virtual (CDC) serial / COM port. The pin diagram in Figure 7.7 and its properties are given in Table 7.4.

It contains everything needed to support the microcontroller. It simply connects to a computer with a micro USB cable to get started. It has a form factor that enables it to be easily placed on a breadboard.

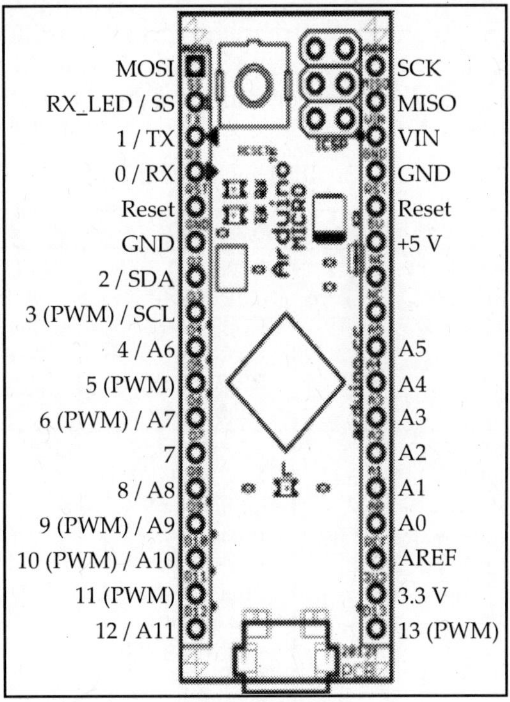

Fig. 7.7 Pin diagram

Table 7.4. Properties of Arduino Micro

Microcontroller	ATmega32u4
Operating voltage	5 V
Digital I/O pins	20
PWM channels	7
Analog input pins	12
DC per I/O pin	40 mA
DC for 3.3V pin	50 mA
Flash memory	32 KB
SRAM	2.5 KB
EEPROM	1 KB
Clock speed	16 MHz

7.3.2 PcDuino

PcDuino is a high performance, cost effective mini PC platform that runs PC like OS such as Ubuntu, Linux. It outputs screen to HDMI enabled TV or monitor via the built-in HDMI interface. It is specially targeted for the fast growing demands from the open source community. The platform could run full blown PC like OS with easy to use tool chain and compatible with the popular Arduino ecosystem such as Arduino Shields, open source projects, etc. Table 7.5 describes various versions of PcDuino with its processor and memory details.

Table 7.5. Description of various versions of PcDuino

PcDuino version	Processor	Ports availability	Memory
PcDuino9s	ARM Cortex A17 RK3288 Quad-core 1.8 GHz	• Micro SD socket (Max 64G) • USB2.0 port OTG port • RS232 port • HDMI port	• 1333 /2 GB DDR3L DRAM • 8G Flash ROM (Max 32 GB)
PcDuino4 Nano	CPU Soc All winner H3 quad core Cortex A7 Mali-400MP2 GPU up to 600 MHz GPU Mali 400 MP2 GPU up to 600 MHz	• 3.5 mm audio jack Via HDMI • Onboard microphone • USB 2.0 × 3 • Microsd slot × 1 • Video output: HDMI 1.4 1080P, CVBS • Debug serial port: 4 Pin, 2.54 mm pitch pin header	DDR3 RAM: 1 GB
PcDuino8 Uno	CPU all winner H8 8-Core Cortex-A7 @ 2.0 GHz GPU Power VR SG544 @720 MHz	• 1 × HDMI 1.4 • CAMERA = MIPI • 1 × USB host • 1 × USB OTG	DDR3 1 GB
PcDuino 3B	CPU all winner A20 SoC, 1 GHz ARM Cortex A7 Dual Core GPC Opengl ES2.0, openvg 1.1, Mali 400 Dual Core	• 3.5 mm analog audio interface • I2S stereo digital audio interface • Camera = CSI • BATTERY = Li-Poly battery interface	• 4 GB Flash • MICROSD card (TF) slot for up to 32 GB • DRAM = DDR3 1 GB

		• 1 × USB host • 1 × USB OTG • HDMI 1.4 with HDCP support	
PcDuino3 Nano	CPU All winner A20 SoC, 1 GHz ARM Cortex A7 Dual Core	• 3.5 mm analog audio interface • 2 × USB host • 1 × USB OTG	• DRAM = 1 GB • 4 GB Flash
	GPU Opengl ES2.0, openvg 1.1, Mali 400 Dual Core		
PcDuino1 Lite Wi-Fi	CPU 1 GHz ARM Cortex A8	• HDMI	• DRAM = 256 MB • 2 GB Flash, microsd card (TF) slot for up to 32 GB
	GPU opengl ES2.0, openvg 1.1 Mali 400 core		
PcDuino1 Lite	CPU 1 GHz ARM Cortex A8	• HDMI	• DRAM = 512 MB • Micro SD card (TF) slot for up to 32 GB
	GPU opengl ES2.0, openvg 1.1 Mali 400 core		
PcDuino 1	CPU 1 GHz ARM Cortex A8	• HDMI	• DRAM = 1 GB • 2 GB Flash, microsd card (TF) slot for up to 32GB
	GPU opengl ES2.0, openvg 1.1 Mali 400 core		
PcDuino 2	CPU 1 GHz ARM Cortex A8	• HDMI	• DRAM = 1 GB • 2 GB Flash, microsd card (TF) slot for up to 32 GB
	GPU opengl ES2.0, openvg 1.1 Mali 400 core		
PcDuino3s	CPU all winner A20 SoC, 1 GHz ARM Cortex A7 Dual Core	• HDMI 1.4 with HDCP support • 3.5 mm analog audio interface • 3 × USB host	• DRAM = 1 GB • 4 GB Flash, microsd card (TF) slot for up to 32 GB
	GPU Opengl ES2.0, openvg 1.1, Mali 400 Dual Core		

Contd.

PcDuino prodto (core board)	CPU all winner A20 soc, 1 GHz ARM Cortex A7 Dual Core GPU Opengl ES2.0, openvg 1.1, Mali 400 Dual Core	• N/A	• DRAM = 1 GB • 4 GB Flash
PcDuino4 STB	CPU all winner H3 Quad-Core Cortex-A7 @ 1.6 GHz H.265/HEVC 4K GPU Mali 400 Mp3 GPU @ 600 MHz, Supports opengl ES2.0	• HDMI 1.4 with HDCP support • 1 × AV output • 2 × USB host	• DRAM = DR3 1 G (shared with GPU) • EMMC/NAND 8G
PcDuino8 STB	CPU all winner H8 8-Core Cortex-A7 @ 2.0 GHz GPU Power VR SG544 @720 MHz	• HDMI 1.4 with HDCP support • 2 × USB host	• DRAM = DDR3 1G • EMMC/NAND 8G

Difference in connectivity, software, programming language among various version of PcDuino is detailed in Table 7.6.

Table 7.6. Description of various versions of PcDuino

PcDuino version	Connectivity	Software	Programming language
PcDuino9s	1REALTEK - TL8211E-VB Gigabit LAN On board 802.11 a/b/g/n/ac, 2.4G On board BT-V4.0 BLE	Ubuntu 14 Android 4.4	• C, C++ with GNU tool chain • Java with standard Android SDK • Python
PcDuino4 Nano	100M Ethernet	Ubuntu 14 Android 4.4	• C, C++ with GNU tool chain • Java with standard Android SDK • Python
PcDuino8 Uno	Gbps RJ45	Ubuntu 14 Android 4.4	• C • PYTHON • JAVASCRIPT

PcDuino 3B	Built-in Wi-Fi Ethernet 10M/ 100 Mbps/ z1 Gbps	Ubuntu 12.04 Android 4.2	• C, C^{++} with GNU tool chain • Java with standard Android SDK • Python
PcDuino3 Nano	Ethernet 10M/100M/1 Gbps	Ubuntu 12.04 Android 4.2	• C, C^{++} with GNU tool chain • Java with standard Android SDK • Python
PcDuino1 Lite Wi-Fi	Wi-Fi	Linux3.0 Ubuntu 12.04	• C, C^{++} with GNU tool chain • Java with standard Android SDK • Python
PcDuino1 Lite	10/100Mbps RJ45 and USB Wi-Fi extension (not included)	Linux3.0 Ubuntu 12.04	• C, C^{++} with GNU tool chain • Java with standard Android SDK • Python
PcDuino 1	10/100 Mbps RJ45 and USB Wi-Fi extension (not included)	Linux3.0 Ubuntu 12.04 Android ICS 4.0	• C, C^{++} with GNU tool chain • Java with standard Android SDK • Python
PcDuino 2	10/100 Mbps RJ45 and USB Wi-Fi extension (not included)	Linux3.0 + Ubuntu 12.04 Android ICS 4.0	• C, C^{++} with GNU tool chain • Java with standard Android SDK • Python
PcDuino3s	Ethernet 10M/100 Mbps	Ubuntu 12.04 Android 4.2	• C, C^{++} with GNU tool chain • Java with standard Android SDK • Python • Scratch for PcDuino • Remote desktop client • Arduino IDE • XBMC

Contd.

PcDuino4 STB	NETWORK 10/100 Mbps Wi-Fi RTL8189ETV	Android 4.4	N/A
PcDuino8 STB	NETWORK 10/100 Mbps Wi-Fi RTL8189ETV	Android 4.4	N/A

7.3.3 Beagle Board

Beagle Board is a low power open source single board computer produced by Texas Instruments in association with Digi-Key and Network element14. The Beagle Board was also designed with open source software development in mind, and as a way of demonstrating the Texas Instrument's OMAP3530 system on a chip. Various versions are listed in Table 7.7 and features with its properties of beagle board are given in Table 7.8.

Table 7.7. Description of various versions of Beagle Board

Beagle Board version	Processor	Memory	Supporting OS
Pocket Beagle	Octavo Systems OSD3358 1 GHz ARM® Cortex-A8	• 512 MB DDR3 RAM integrated	• Linux
Beagle Bone Blue	Octavo OSD3358 AM335× 1 GHz ARM® Cortex-A8 processor	• 512 MB DDR3 RAM • 4 GB 8-bit eMMC flash storage	• Linux
Beagle Bone Black	AM335× 1GHz ARM® Cortex-A8	• 512 MB DDR3 RAM • 4 GB 8-bit eMMC on-board flash storage	• Debian • Android • Ubuntu • Cloud9 IDE on Node.js • w/ Bone Script library
Beagle Bone AI	Texas Instruments Sitara AM5729	• 2.5 MB of on-chip L3 RAM • 1 GB RAM and 16 GB on-board eMMC flash with high-speed interface	• Linux

BeagleBone® Black Wireless	Octavo Systems OSD3358 1 GHz ARM® Cortex-A8	• 512 MB DDR3 RAM • 4 GB 8-bit eMMC on-board flash storage	• Debian with Cloud9 IDE on Node.js w/ Bone Script library • Third party support for Android and Ubuntu
Seeed Studio Beagle Bone Green	AM335× 1 GHz ARM® Cortex-A8	• 512MB DDR3 RAM • 4 GB 8-bit eMMC on-board flash storage	• Debian • Android • Ubuntu • Cloud9 IDE on Node.js w/ BoneScript library
Seeed Studio Beagle Bone Green Wireless	AM335× 1 GHz ARM® Cortex-A8	• 512 MB DDR3 RAM • 4 GB 8-bit eMMC on-board flash storage	• Debian • Android • Ubuntu • Cloud9 IDE on Node.js
San Cloud Beagle Bone Enhanced	AM335× 1 GHz ARM® Cortex-A8	• 1 GB DDR3 RAM • 4 GB 8-bit eMMC on-board flash storage	• Debian • Android • Ubuntu • Cloud9 IDE on Node.js w/ BoneScript library
Beagle Bone	AM335× 720 MHz ARM Cortex-A8	• 256 MB DDR2 RAM	• 4 GB microSD card w/ Angstrom Distribution • Cloud9 IDE on Node.JS w/ BoneScript library
BeagleBoard-X15	TI AM5728 2×1.5 GHz ARM® Cortex-A15	• 2 GB DDR3 RAM • 4 GB 8-bit eMMC on-board flash storage	• Debian • Android • Ubuntu • Cloud9 IDE on Node.js
Beagle Board-Xm	AM37× 1 GHz ARM Cortex-A8 compatible	• 512 MB LPDDR RAM	• Angstrom Linux • Android • Ubuntu • XBMC

Table 7.8. Features of various versions of Beagle board

Beagle board version	Image	Features
Pocket Beagle		• 8 analog inputs, • 44 digital I/Os • MicroUSB host/client • MicroSD connectors
Beagle Bone Blue		• Wireless: 802.11bgn, Bluetooth 4.1 and BLE • Motor control: 8 6 V servo out, 4 bidirectional DC motor out, 4 quadrature encoder in • Sensors: 9 axis IMU (accels, gyros, magnetometer), barometer, thermometer • User interface: 11 user programmable LEDs, 2 user programmable buttons
Beagle Bone Black	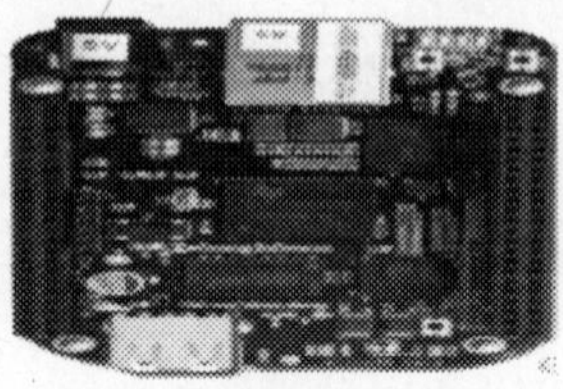	• USB client for power & communications • USB host • Ethernet • HDMI • 2 × 46 pin headers
Beagle Bone AI		• USB type-C for power and superspeed dual-role controller; and USB type-A host • Gigabit Ethernet, 2.4/5 GHz Wi-Fi, and Bluetooth • microHDM
Beagle Bone Black Wireless		• USB client for power & communications • USB host • 802.11b/g/n and Bluetooth 4.1 plus BLE • HDMI • 2 × 46 pin headers

<table>
<tr><td>Seed Studio Beagle Bone Green</td><td></td><td>• USB client for power & communications
• USB host
• Ethernet
• 2 × 46 pin headers
• 2 × Grove connectors (I2C and UART)</td></tr>
<tr><td>Seed Studio Beagle Bone Green Wireless</td><td>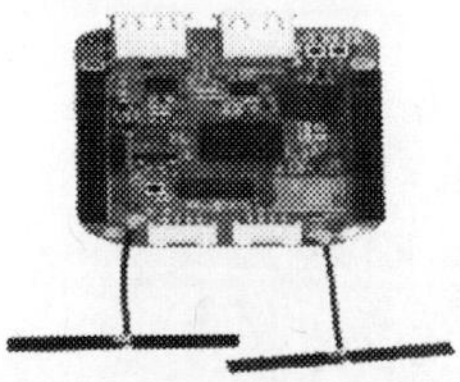</td><td>• USB client for power & communications
• USB host with 4-port hub
• Wi-Fi 802.11 b/g/n 2.4 GHz
• Bluetooth 4.1 with BLE
• 2 × 46 pin headers
• 2 × Grove connectors (I2C and UART)</td></tr>
<tr><td>San Cloud Beagle Bone Enhanced</td><td></td><td>• USB client for power & communications
• 4 × USB host
• Gigabit Ethernet
• 2 × 46 pin headers</td></tr>
<tr><td>Beagle Bone</td><td></td><td>• USB client: power, debug and device
• USB host
• Ethernet
• 2 × 46 pin headers</td></tr>
<tr><td>BeagleBoard-X15</td><td></td><td>• 2 × Gigabit Ethernet
• 3 × Super Speed USB 3.0 host
• High Speed USB 2.0 client
• eSATA (500 mA)
• full-size HDMI video output
• microSD card slot
• Stereo audio in and out
• 4 × 60-pin headers with PCIe, LCD, mSATA</td></tr>
<tr><td>BeagleBoard-xM</td><td>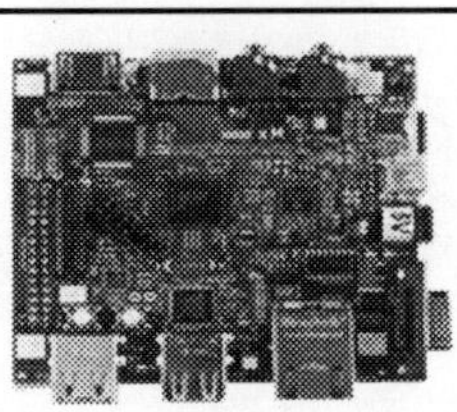</td><td>• 2D/3D graphics accelerator
• 4 USB 2.0 ports
• MMC/SD connector
• DVI-D port
• S-Video port
• USB mini AB connector
• Ethernet</td></tr>
</table>

7.3.4 Cubie Board

Cubie Board is a complete computer built on a single circuit board, with microprocessor, memory, input/output and other features required of a functional computer. Single board computers are commonly made as demonstration or development systems, for educational systems and for use as embedded computer controllers. Various versions of Cubie Board with its properties are given in Figure 7.8 to 7.12 and Table 7.9 to 7.13.

(a) CubieAIO-A20 Board

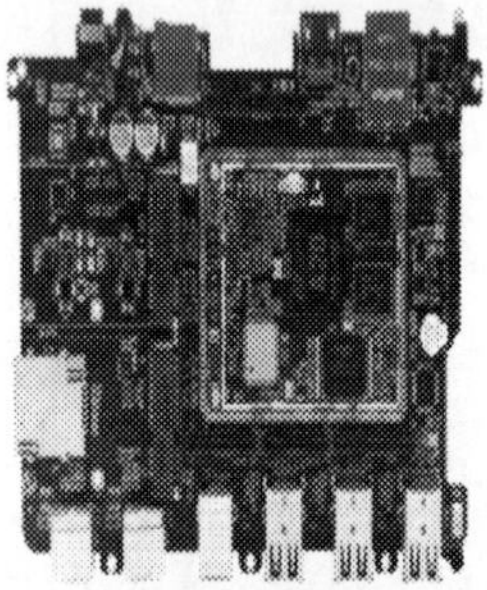

Fig. 7.8 Pin diagram

Table 7.9. Properties and features

Processor	All winner A20 A7 Dual Core Mali 400 MP2
Memory	• 1 GB DDR3 • 8-bit width • 256 MB*2
Network Interface	• GIGABIT RJ245 • Wi-Fi module
Features	• HDMI 1.4A • UART × 4 • USB2.0 TYPE A × 6 • Support capacitor touch screen • Simslot available support 3G/4G

(b) Cubieboard 5

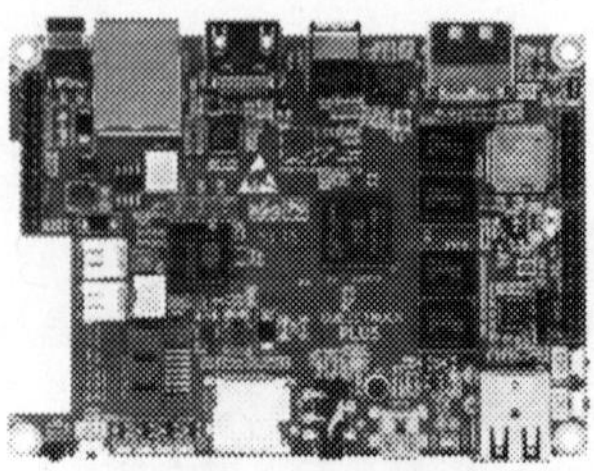

Fig. 7.9 Pin diagram

Table 7.10. Properties and features

Processor	All Winner SOC H8, ARM® Cortex-A7 Octa-Core
Memory	• 1 GB/2 GB DDR3@480 MHz (960 MTPS), 2 GB Default
Network Interface	• GIGABIT RJ245 • Wi-Fi + BT4.0 wireless connection with an-enna on-board
Features	• 2 × USB HOST • 1 × OTG, • 1 × Toslink (SPDIF Optical)

(c) Cubie Board 4

Fig. 7.10 Pin diagram

Table 7.11. Properties and features

Processor	All winner A80 28 nm Octa-Core A15/A7 big
Memory	• 2 GB DDR3
Network Interface	• GIGABIT RJ245
Features	• 4 *USB2.0 host ports • 1 *USB3.0 OTG port • Audio output, • Microphone input

(d) Cubie Board 3

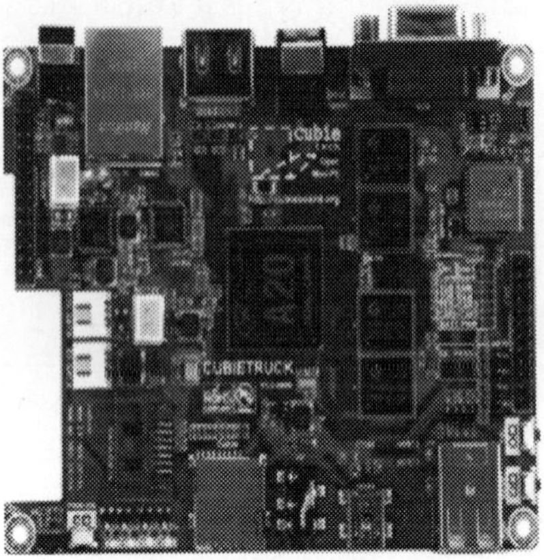

Fig. 7.11 Pin diagram

Table 7.12. Properties and features

Processor	All winner technology A20, Arm architecture Cortex-A7 dual core
Memory	• 1 GB/2 GB DDR3@480 MHz (960 MTPS)
Network Interface	• GIGABIT RJ245
Features	• 2 USB host ports • 1 USB OTG port • 1 earphone port • a micro SD card slot, • VGA and HDMI display port, • optical SPDIF port for HD audio

(e) Cubie Board 2

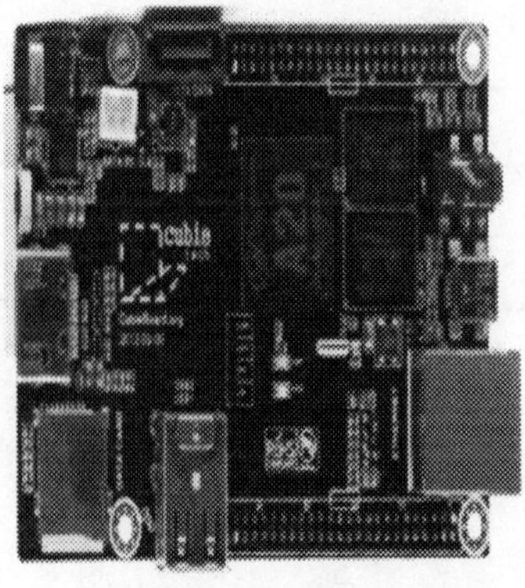

Fig. 7.12 Pin diagram

Table 7.13. Properties and features

Processor	All Winner SOC A20, ARM® Cortex-A7 Dual-Core
Memory	• 1 GB DDR3 @ 480 MHz (960 MTPS) • eMMC, 4 GB default • 8 GB is an option • Micro SD card slot up to 32 GB
Network Interface	• GIGABIT RJ245(10M/100M)
Features	• HDMI Port A • HDMI V1.4a

Cubie Board 1 has the same design as Cubie Board 2.

7.3.5 Contiki OS and COOJA Simulator

Contiki is an open source OS like Windows and Linux. It can be used only for a specific purpose and mainly focused on 'things' that are going to automate in the IoT. In general, OS performs certain functions they are program/process management, resource management, memory management and communication management.

The goal of Contiki is mainly to meet the requirements of the smallest things used in IoT, that is, things should be able to communicate a few bits of data to one another. Contiki has a vision to create minimized things working OS for wide deployment area. This OS has broken many myths about the smallest footprint in which an OS can be stored and made to function. It is integrated with the smallest IPV6 stack, i.e., μIPV6 stack. Contiki provides high level completeness and flexibility to the programmers which make it different from other small tiny OS.

(a) Features and functions of Contiki

Contiki is completely an open-source software. It comes with a rich set of features that are device and programmer friendly. This software has consumers in companies like Cisco and Google. It was distributed using the 3clause BSD style license. The integrity of the software is ensured by running the nightly regression tests.

Contiki OS can fit into the devices with 10 KB of RAM and 100 KB of ROM. It can run on devices such as 8051 SoC to ARM powered devices. This software enabling ports are available on other working platforms such as Arduino and Atmel. To interoperate with public networks (such as Internet) standard compliance could be needed with more efficient and optimized protocols. One such example is the Rime stack, developed to optimize the IPV6 protocol.

Contiki comes with much documentation apart from well documented code. Codes for performing various applications are available in github. Various functions of Contiki OS are given as follows.

1. **Process and memory management:** Contiki supports memory block allocation and standard 'C' style allocation with malloc(). The concepts and implementations of 'protothreads' are introduced to keep the low system requirements. 'protothreads' are C-language based implementation minimizing the overhead of multi-threaded programming.
2. **Communication management:** Contiki supports both IPV4 and IPV6 stack implementations, which include TCP, UDP and HTTP protocols with the smallest footprints. The IP stack is standards compliant. The OS also supports low power variants such as 6LoWPAN.
3. **File system management:** Not all devices as part of IoT have luxury of persistent storage such as Flash. But for such devices, Contiki provides support through the Coffee Flash file system.

(b) Applications

There are several applications that come packaged as part of Contiki. These include a small Web browser, Web server, Calculator, Shell, Telnet Client and Daemon, Email client, VNC viewer and FTP.

Developers can find tools such as Cooja simulator for application development which will be explained in next section. Contiki also provides mechanisms to measure power for various operations that help developers produce applications that are power sensitive. Software bundle also comes with a UNIX style shell for interacting with OS and debugging.

(c) Communication components in Contiki (μIP and μIPV6)

μIP: There are several advantages of having the Internet protocol stack embedded in the OS. It allows communication with the existing devices/ computers in the Internet. For this reason, the IP stack has been trimmed to fit 1 KB of RAM and a few KB of ROM, in comparison with Linux based IP stack that requires 1 MB RAM. This trimming of memory comes at the cost of reduced throughput.

μIPV6: It is in contrast to the IPV6 certified version of IPV6, which was developed by Cisco. Support of IP stack along with TCP and UDP protocols makes it possible for devices running Contiki to directly interact with a Web server or tweak messages to the Internet.

LoWPAN (IEEE 802.15.4) and 6LoWPAN (Low Power Wireless Personal Area Networks) have the characteristics of small packet sizes, low data rates, low power devices and large number of devices. The IPv6 extension of the same is the 6LoWPAN standard developed by IETF working group. For saving power, the transmission and receiving periods are short spaced along with long periods of sleep mode by devices. Also, for power constraint reasons the communication is made hop based and short range.

(d) COOJA Simulator

This section gives a high level overview of sensor network simulations using COOJA. It may help users to more easily understand problems, limitations, and features in the simulator.

Contiki level: the Contiki Mote Type

A simulated Contiki Mote using COOJA is used to control and analyze the sensor connected environment. It uses native platform as a shared library and loading the library into Java using Java Native Interfaces (JNI). Several different Contiki libraries can be compiled and loaded in the same COOJA simulation, representing different kinds of sensor nodes (heterogeneous networks). COOJA controls and analyzes a Contiki system via a few functions. For instance, the simulator informs the Contiki system to handle an event, or fetches the entire Contiki system memory for analysis. This approach gives the simulator full control of simulated systems. Unfortunately, using JNI also has some annoying side-effects. The most significant is the dependency on external tools such as compilers and linkers and their runtime arguments. COOJA was originally developed for Cygwin/Windows and Linux platform, but has later been ported to MacOS.

7.4 RASPBERRY PI INTERFACES

(a) Introduction

In order to get smoother, faster and more comfortable lifestyle, connectivity between device and data processing is the technological demand. With low cost data storage, processing and communication between gadgets, day-to-day activities have to be automated and connected at the same time. Internet of Things (IoT) based applications are possible at low cost, ease of access and flexibility to use by Raspberry Pi. Raspberry Pi with Sensors connected to it provides the things with Internet Gateway. Raspberry Pi is the next generation computing device with a low cost and capacity of functioning as a computer. With this small device, one can design and realize several applications and prototype models with the minimum knowledge of programming.

The Raspberry Pi is a powerful, small-single board computer. This computer uses ARM (Advanced RISC Machine) processor, the processor at the heart of the Raspberry Pi system is a Broadcom BCM2835 System on Chip (SoC) multimedia processor. It uses a many different kinds of processors. We can install several third party versions of OS like Ubuntu, Archlinux, RISC OS, Windows 10 IoT Core, etc. By connecting peripherals like keyboard, mouse, display to the Raspberry Pi, it will act as a mini computer. Raspberry Pi is also used to surf the Internet, to send an email to write a letter using a word processor and can do much more. It is slower than laptop or desktop but is still a computer which can provide all the expected features or abilities, at low power consumption. Raspberry Pi is affordable and in addition difficult to break.

(b) Features of Raspberry Pi

- The Raspberry Pi is a small independent computer that runs on the various distributions of Linux operating system and can be programmed as needed.
- It provides access to the on-chip hardware, i.e., GPIOs for developing an application. By assessing GPIO can connect devices like LED, motors, sensors, etc., and can control them too.
- Operating speed of Raspberry Pi is from 700 MHz to 1.2 GHz.
- It supports USB 2.0 which is used to connect various peripherals.
- It has large working memory and expandable memory to store data.
- It has on-board SDRAM that ranges from 256 MB to 1 GB.
- It is possible to expand the Raspberry Pi with Wi-Fi and Bluetooth adaptors.
- Raspberry Pi supports various sensors and electronic components using digital input and output.
- It also provides on-chip SPI, I2C, I2S and UART modules.
- C, C^{++}, Python, Java can be used for programming of Raspberry Pi.

(c) Architecture of Raspberry Pi 4

The Raspberry Pi 4 Model B (Pi4B) is the new generation of Raspberry Pi computers supporting more RAM and enhanced CPU, GPU and I/O performance The Pi4B is available with 1, 2 and 4 gigabytes of LPDDR4 SDRAM. The architectural diagram is given in Figure 7.13.

Features

Hardware

- Quad core 64-bit ARM-Cortex A72 running at 1.5 GHz
- 1, 2 and 4 gigabyte LPDDR4 RAM options
- H.265 (HEVC) hardware decode (up to 4Kp60)

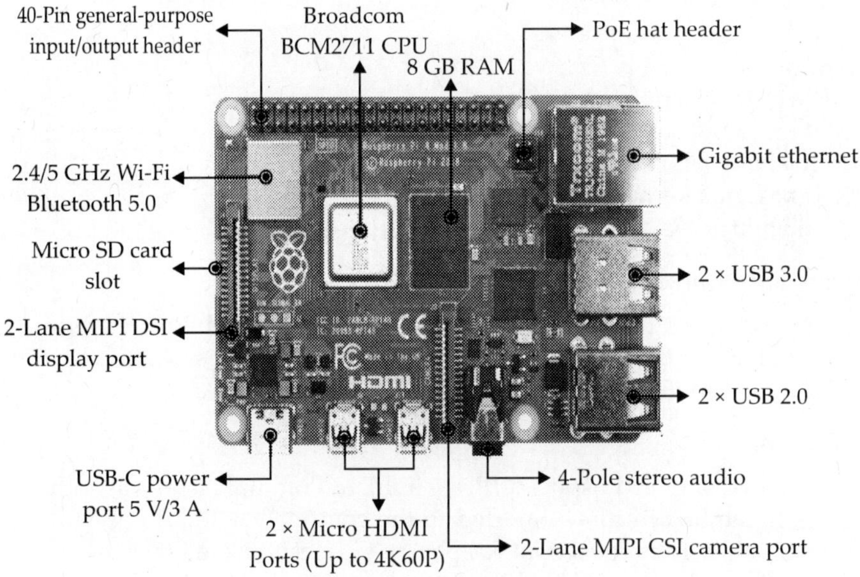

Fig. 7.13 Architecture of Raspberry Pi

(https://blog.seeedstudio.com/wp-content/uploads/2020/05/44671590483676_.pic_hd-1030x611.png)

- H.264 hardware decode (up to 1080p60)
- VideoCore VI 3D Graphics
- Supports dual HDMI display output up to 4Kp60

Interfaces

- 802.11 b/g/n/ac Wireless LAN
- Bluetooth 5.0 with BLE
- 1 × SD Card
- 2 × micro-HDMI ports supporting dual displays up to 4Kp60 resolution
- 2 × USB2 ports
- 2 × USB3 ports
- 1 × Gigabit Ethernet port (supports PoE with add-on PoE HAT)
- 1 × Raspberry Pi camera port (2-lane MIPI CSI)
- 1 × Raspberry Pi display port (2-lane MIPI DSI)
- 28 × user GPIO supporting various interface options:
- Up to 6 × UART
- Up to 6 × I2C
- Up to 5 × SPI
- 1 × SDIO interface
- 1 × DPI (Parallel RGB Display)
- 1 × PCM

- Up to 2 × PWM channels
- Up to 3 × GPCLK outputs

Software

- ARMv8 Instruction Set
- Mature Linux software stack
- Actively developed and maintained
- Recent Linux kernel support
- Availability of GPU functions using standard APIs

Power Requirements

The Pi4B requires a good quality USB C power supply capable of delivering 5 V at 3 A. If attached downstream USB devices consume less than 500 mA, a 5 V, 2.5 A supply may be used.

Connection

The Ethernet port in Raspberry Pi is used to communicate with other devices. It can be directly connected to the computer without need for a crossover cable. The model A of Raspberry Pi doesn't have Ethernet port, so it is connected through USB Ethernet adapter having two speed modes, 10 Mb/s and 100 Mb/s. By using cable, Raspberry Pi automatically receives the details Dynamic Host Configuration Protocol (DHCP). The model B has a standard RJ45 Ethernet port.

Operating Systems and Programming Languages

Raspberry Pi is a board which provides Debian based Raspbian OS. Also, they provide NOOBS (New Out Of the Box Software) OS for Raspberry Pi. Features of Raspberry Pi include amazing HD (high definition) quality, video playback, high quality audio and has the capability to play 3D games. Software such as RASPBIAN, PIDORA, OPENELEC, RASPBMEC, RISC OS, and ARCH LINUX are used. This software can be downloaded easily and are free from the official forum under the NOOBS category. Raspberry Pi supports Python as the main programming language for functioning and coding. It also supports BASIC, C, C^{++}, JAVA, and Perl and Ruby languages.

Disadvantages of Raspberry Pi

- It does not have a Basic Input Output System (BIOS). So, it always boots from an SD card.
- It does not have Analog to Digital converter and Real Time Clock (RTC).

(d) Raspberry Pi based IoT Applications

IoT devices are low powered SoC devices with a microcontroller as the core processor and a Wi-Fi module as the communication interface. IoT primarily connects devices (things) on the Internet for monitoring and control in real time. If the PC is used as an IoT device, it requires main powers of many hundreds of watts. Raspberry Pi is capable of providing Internet gateway as it has ARM processor. The raspberry Pi is very small and uses very little energy. Raspberry Pi board is designed for open-source development.

Raspberry 2 model B and raspberry 3 model B and higher versions are preferred to construct sensor node. Raspberry 2 model B typically consumes power less than 2 W when conducting scientific computing. With a Broadcom BCM2836 system on chip, which consists of a 900 MHz quad-core ARM Cortex-A7 CPU and 1 GB RAM, Raspberry Pi is capable of complex operation. Raspberry Pi has 4 USB ports, 40 GPIO pins, full HDMI port, Ethernet port, display interface (DSI), micro SD card slot, and 3D graphics core, which provide more possibilities for its application. Recently, MATLAB launched its package and open-source toolboxes on Raspbian. Therefore, MATLAB based algorithms can also be implemented on Raspberry Pi. RS485 bus is used to connect all sensors. MODBUS protocol is applied on Raspberry Pi to start sensor and get data through a USB-RS485 converter. Raspberry Pi stores the received data and then processed. After calculation, Raspberry Pi sends results to local registered clients through Wi-Fi (wireless fidelity) communication module and database server will receive data package through GPRS (General Packet Radio Service) module. The Raspberry Pi 3 comes with on board Wireless LAN (WLAN 802.11n), i.e., Wi-Fi and Bluetooth adapters. This means that all need is Raspberry Pi 3 for getting connected to Wi-Fi or Bluetooth and there is no need for additional peripherals like USB Dongles (for Wi-Fi or Bluetooth).

(e) Different Versions of Raspberry Pi

Raspberry Pi is developed by Broadcom in collaboration with Raspberry Pi Foundation. All generations of Raspberry Pi having a Broadcom system on a chip with an integrated ARM compatible central processing unit (CPU) and on chip graphics processing unit (GPU).

- The first generation (Raspberry Pi Model B) was released in February 2012, followed by the simpler and cheaper Model A. The improved design is released in 2014, which is Raspberry Pi Model B+. These boards are approximately credit card sized. Improved A+ and B+ models were released a year later.
- The Raspberry Pi 2, which featured a 900 MHz quad core ARM Cortex A7 processor and 1 GB RAM, was released in February 2015.
- In November 2015, a Raspberry Pi Zero with smaller size and reduced input/output (I/O) and general purpose input/output (GPIO) capabilities

was released. On 28 February 2017, the Raspberry Pi Zero W was launched, a version of the Zero with Wi-Fi and Bluetooth capabilities.

- On 12 January 2018, the Raspberry Pi Zero WH was launched, a version of the Zero W with pre-soldered GPIO headers.
- Raspberry Pi 3 Model B was released in February 2016 with a 1.2 GHz 64bit quad core processor, on board 802.11n Wi-Fi, Bluetooth and USB boot capabilities.
- The Raspberry Pi 3 Model B+ was launched with a faster 1.4 GHz processor and a three times faster gigabit Ethernet (throughput limited to 300 Mbit/s by the internal USB 2.0 connection) or 2.4 / 5 GHz dual band 802.11ac Wi-Fi (100 Mbit/s). Other features are Power over Ethernet (PoE) (with the add on PoE HAT), USB boot and network boot (an SD card is no longer required).
- Raspberry Pi 4 Model B was released in June 2019 with a 1.5 GHz 64-bit quad core ARM Cortex-A72 processor, on board 802.11ac Wi-Fi, Bluetooth 5, full gigabit Ethernet (throughput not limited), two USB 2.0 ports, two USB 3.0 ports, and dual monitor support via a pair of micro HDMI (HDMI Type D) ports for up to 4K resolutions.

(f) Comparison with Arduino

Arduino is an open source physical computing platform based on a simple microcontroller board and a development environment for writing software for the board. It can receive input from a variety of sensors and can affect its surroundings by controlling lights, motors, and other actuators. The microcontroller on the hardware board can be programmed using Arduino programming language and the Arduino Integrated Development Environment (IDE). Arduino supports two working modes, stand-alone or connected to a computer via USB cable. Arduino works with X86 processors and Raspberry Pi works with SoC with ARM Cortex processor as its CPU. X86 processor uses Complex Instruction Set Computer (CISC) and requires powerful platform and complex hardware. ARM processor works on Reduced Instruction Set Computer (RISC). Its efficiency is high and low power is consumed. X86 processor has less efficiency than ARM processor.

The Raspberry Pi is faster than an Arduino when it comes to clock speed. The Raspberry Pi is an independent computer that can run an actual operating system in Linux. It can multitask, support two USB ports, and connect wirelessly to the Internet. It can run multiple programs in the background while activated.

The most important features when it comes to building an IoT system using Arduino or Raspberry Pi boards is connectivity capabilities. The latest Raspberry Pi models have Wi-Fi, Bluetooth, and Ethernet connectivity installed by default, offer a pretty good speed and processing power which can handle video and audio data. Arduino requires modules to add basic connectivity options like Wi-Fi, GPS or Bluetooth.

When comparing Arduino board with Raspberry Pi for a simple single action IoT application, memory storage may not be the most important factor. However, if your IoT device is multifunctional, it may require substantial RAM. As a microcontroller, Arduino UNO has only 32K bytes of flash memory and 2K bytes of SRAM. Raspberry Pi has much larger memory capacity, the latest models are said to provide up to 8 GB of SDRAM. It's enough to run multiple functionalities for a more complex IoT system.

Raspberry Pi can handle multitasking and run more complex functionality than Arduino. It includes running media playback, making calculations and collecting various parameters at the same time. So if an IoT system needs to collect data from several sensors, pull data from the Internet, connect to a smart phone and provide a complex output on a display, Raspberry Pi is best suited for IoT applications compared to Arduino.

While comparing the overall performance, computing power, size and cost of Raspberry Pi with other popular boards and development platforms, Raspberry Pi is ultra cheap board, with large number of input and output peripherals. It is the perfect platform for interfacing with a wide variety of external peripherals and has wide range of applications. Raspberry Pi provides access to Internet by coupling it with Wi-Fi. It is best suited for IoT applications by providing access to sensor networks. Thus, the advantage of Raspberry Pi lies in its flexibility that user can program it according to their needs and budget.

7.5 CLOUD OFFERINGS OF IoT

7.5.1 Cisco and IBM IoT Platforms

IBM and Cisco has announced a global collaboration to provide Internet of Things at the edge of the network. Together, they are enabling businesses and organizations in remote and autonomous locations to tap into the combined power of IBM's Watson IoT Cloud and business analytics technologies and Cisco's edge and fog capabilities to more deeply understand and act on critical data on the network edge. It allows customers to analyze business performance at the point of data collection, monitor and control how their environment, assets and people are performing against their operation goal. In this audio and video transmissions are carefully monitored to avoid security breach. Frequent analytic evaluations done based on industrial business standard. Analysis at the edge reduces more number of pushing data to Cloud, increases data transmission capacity and reducing communication costs.

It provides an advantage of tapping previously connected data and empowering decisions at the point of data collection. It enables dynamic distribution of data processing and analytics with performance critical computation for application functions at the very edge of the network. As an answer to define edge analytics multiple controlling devices connected to the local gateway to analyze and process data before sending to the Cloud. In some cases cognitive IoT with in-built artificial intelligence to suit real-time applications.

7.5.2 RESTful web API

REST is acronym for REpresentational State Transfer. A RESTful API is an architectural style for an Application Program Interface (API) that uses HTTP requests to access and use data. Data can be used to GET, PUT, POST and DELETE data types, which refers to the reading, updating, creating and deleting of operations concerning resources. It was first presented by Roy Fielding in 2000.

An API is a code which allows the communication between software programs. It provides the proper way for a developer to write a program requesting services from an operating system or other application. It was used by Cloud consumers to expose and organize access to web services.

REST technology is generally preferred over other similar technologies because it uses less bandwidth, efficient Internet usage. RESTful APIs can also be built with programming languages such as JavaScript or Python. REST is a logical choice for building APIs that allow users to connect, to manage and to interact with Cloud services flexibly in a distributed environment. RESTful APIs are used by some sites as Amazon, Google, LinkedIn and Twitter.

How RESTful APIs work?

A RESTful API breaks down a large transaction to series of small modules. Each module addresses an underlying part of the transaction. This modularity provides developers with a lot of flexibility, but it can be challenging for developers to design their REST API from scratch. Companies like Amazon S3, Cloud Data Management Interface (CDMI) and Open Stack Swift are the most popular to provide this service. A RESTful API uses commands to obtain resources. The state of a resource at any given timestamp is called a resource representation. A RESTful API uses existing HTTP methodologies defined by the RFC 2616 protocol, such as:

- GET to retrieve a resource
- PUT to change the state of or update a resource, which can be an object, file or block
- POST to create that resource
- DELETE to remove it.

Data formats the REST API supports include:

- application/json
- application/xml
- application/x-wbe+xml
- application/x-www-form-urlencoded
- multipart/form-data

Uses

The call in the rests is stateless, so it is suitable for Cloud applications. Stateless components can be freely redeployed if something fails, and they can scale to accommodate load changes. This is because any request can be directed to any instance of a component; there can be nothing saved that has to be remembered by the next transaction. That makes REST preferable for web use because it provides a service of URL decoding.

7.5.3 AMAZON Web Services

AWS (Amazon Web Services) is a comprehensive, evolving Cloud computing platform provided by Amazon. It provides three important services of Cloud. They are infrastructure as a service (IaaS), platform as a service (PaaS) and packaged software as a service (SaaS) offerings. AWS services offers certain tools to compute power, database storage and content delivery services. AWS was launched in 2006 as Amazon. com built to handle its online retail operations. It was the first company to introduce a pay-as-you-go Cloud computing model. AWS offers many different tools and solutions for enterprises and software developers. It can be used in data centers in 190 countries. Groups such as government agencies, education institutions, non-profit and private organizations can use AWS services.

AWS in IoT

AWS also has a variety of services that enable the Internet of Things (IoT) deployments. The AWS IoT service provides a backend platform to manage IoT devices and data ingestion to other AWS storage and database services. The AWS IoT Button provides hardware for limited IoT functionality and AWS Green grass brings AWS compute capabilities to IoT devices.

Working of AWS

AWS is separated into different services; each can be configured in different ways based on the user's needs. Users should be able to see configuration options and individual server maps for an AWS service. Its services include Compute, Storage databases, Data management, Migration, Hybrid Cloud, Networking, Development tools, Management, Monitoring, Security, Governance, Big data management, Analytics, Artificial intelligence (AI), Mobile development, Messages and notification.

7.6 DATA ANALYTICS FOR IOT

Data analytics is a science of analysis in which statistical techniques are applied to big and small datasets with varying data properties to extract meaningful conclusions and actionable insights for making smart decisions from the uncovered hidden patterns, unknown correlations, trends and any other useful business information. It is a technique used by data scientists and analysts to make smart decisions concerned with their research and businesses. The smart solutions are based on graphical response, bulk data and specifications obtained from earlier works. Hence, data analytics is defined as the process of collecting raw data which are then transformed into meaningful information that will help the user to make decisions on the basis of key insights drawn from the analysis. The core information is brought to the forefront in data analytics these will provide an easy to understand metrics for the user. Data analytics helps companies to better understand their customers, evaluate their ad campaigns, personalize content, create content strategies and develop products.

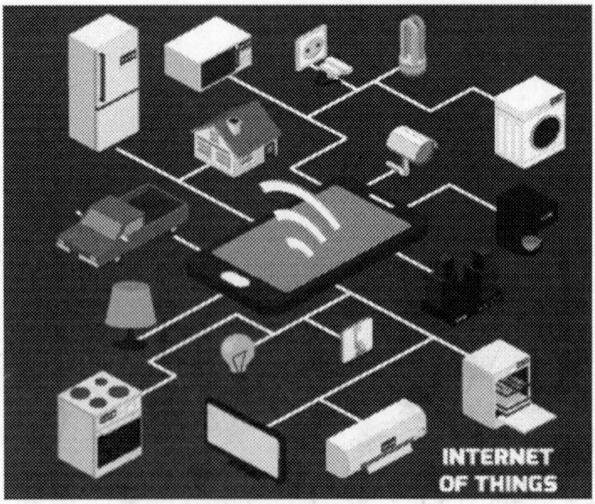

Fig. 7.14 Mobile based smart things

(www.fingnet.com)

Every device, sensor, software in a particular area is connected to each other through the Internet up to a certain distance. This interconnection between things that forms a complete system can be monitored with smart devices such as mobiles, watches, tablets or laptops. The power to access the different things through smart devices is called the Internet of Things (IoT). The smart devices form an important component in IoT. The smart devices generate a massive amount of data which need to be explored and investigated in real time. The data generated from IoT devices is value only when it is subjected to analysis. Hence, need to go in for data analytics.

Embedded systems are networked to communicate, collaborate, correlate and corroborate with one another. With the maturity of the Internet of Things (IoT) technologies and cyber physical systems (CPSs), Device to Device (D2D) and Device to Cloud (D2C) integration thrive towards sophisticated applications. The terms Digitization (IoT) and Digital Intelligence (IoT Data Analytics / Machine Learning) have paved the way for digitally empowered and transformed enterprises. The combination of IoT and data analytics has already found applications in the fields of retail, healthcare, telemetric, manufacturing, and smart cities.

(a) Why IoT Data Analytics?

1. Smarter environments are established (smarter homes, hotels, hospitals, etc.).
2. Timely and actionable insights for both machines and men are obtained.
3. Realization of smart objects, devices, networks and environments is enabled.
4. Pioneering and people-centric applications and services are produced.
5. Precise predictions and prescriptions can be altered.
6. Process excellence and people productivity is facilitated.
7. Preventive maintenance of infrastructures is guaranteed.
8. Optimized utilization of distributed assets is ensured through monitoring, measurement, and management for perfect inventory replenishment.
9. Safety and security of people and properties are safeguarded.
10. Complex environments are monitored to guarantee business performance, productivity and resilience.

(b) Why IoT Data Analytics on Clouds?

- **Agility and affordability:** The analytical workloads do not require any capital investment of large size infrastructures. It can just be used and paid later. When the need goes down they are quickly provisioned and decommissioned.

- **Data Analytics Platforms in Clouds:** Cloud enabled and ready platforms like generic or specific, open or commercial grade, etc., are fast and easy to be leveraged.
- **NoSQL and NewSQL Databases and Data Warehouses in Clouds:** The next generation data analytics is speeded up in Cloud using all kinds of database management systems and data warehouses. The rapid realization of Analytics as a Service (AaaS) is provided by Database as a Service (DaaS), Data Warehouse as a Service (DWaaS), Business process as a Service (BPaaS) and other advancements.
- **WAN Optimization Technologies:** Large quantities of data can be transmitted quickly over the Internet infrastructure using WAN optimization products.
- **Social and professional networking sites:** It uses public Cloud environments.
- **Enterprise class Applications in Clouds:** Applications of all kinds which are customer facing are Cloud enabled and deployed in highly optimized and organized Cloud environments.
- **Anytime, anywhere, any network and any device information and service access:** It can be activated through Cloud based deployment and delivery.
- **Cloud Integrators, Brokers and Orchestrators:** Products and platforms are available to provide seamless interoperability among geographically distributed Cloud environments. There are collaborative efforts towards federated Clouds and the Inter Cloud.
- Sensor/Device-to-Cloud Integration Frameworks are available for transmitting ground level data to Cloud storage and processing.

Distinct Capabilities of IoT Data Analytics Platforms

1. Scalability
2. Faster data ingestion
3. Better read and write performance
4. Faster query processing
5. Portability (the ability to run in edge, private and public clouds) and flexibility.
6. Distributed processing through automated sharing
7. Better data compression
8. Provides an integrated and end to end platform for all kinds of data and analytics
9. Machine and deep learning capabilities
10. RESTful interfaces
11. In-Memory and In-Database Analytics

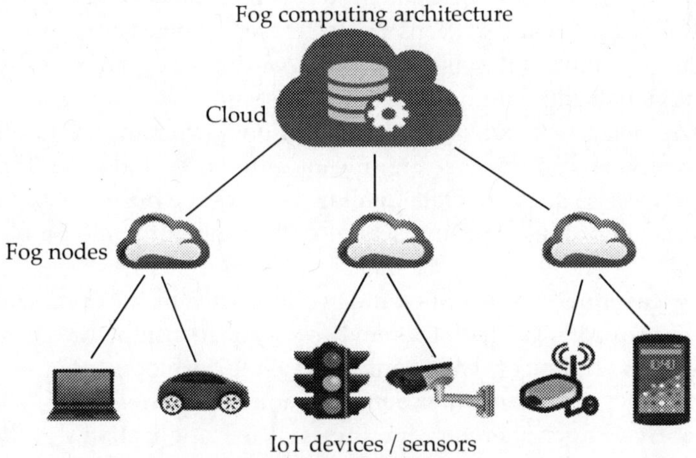

Fig. 7.15 IoT edge analytics

The two major approaches for IoT edge analytics are:

1. The first approach is to keep a special appliance or clusters of appliances embedded with IoT edge analytics platforms and SDKs in a corner of the environment.
2. The second approach is to form an ad hoc edge device Cloud by combining the resource-intensive devices in the environment and deploy the IoT data analytics platform to capture, stock and process digital data in real time and at scale as required.

The participating devices could be stuffed with edge analytics software to be smart in their offerings, operations and outputs. A simple edge node is shown in Figure 7.15.

(c) IoT Edge Data Analytics Appliances and Platforms

1. Dell Edge Gateways for IoT
2. HPE Edgeline IoT Systems
3. IBM Watson at the Edge (Cognitive Edge Analytics)
4. AXON the IoT Platform
5. GE Predix Platform
6. FogHorn
7. The Azure IoT Gateway SDK

(d) Role of Data Analytics in IoT

Data analytics plays a significant role in the growth and success of IoT applications and investments. Analytics tools allow the business units to make effective use of their datasets.

- **Volume:** Large volume of data needs to be managed and analyzed for extracting relevant patterns by IoT based business organizations. These large volume datasets along with real-time data can be easily and efficiently analyzed with data analytics software.
- **Structure:** Unstructured, semi-structured and structured datasets are involved in IoT applications. Significant difference in the data formats and types may also exist. Data analytics allows the business executive to analyze all of these varying sets of data using automated tools and software.
- **Driving Revenue:** The revenues and profits earned by the organizations can be improved by using data analytics to gain an insight into customer preferences and choices. This would lead to the development of services and offers as per the customer demands and expectations.
- **Competitive Edge:** There are a number of IoT application developers and providers in the market but with the use of data analytics in IoT investments better services can be provided by a business unit which helps to gain a competitive edge in the market.

(e) Challenges of IoT Data Analytics

Data analytics involves a big data processing in which a large amount of data need to be analyzed on a real-time basis and stored using multiple storage techniques. The steps that are considered for data processing are:

1. A massive amount of heterogeneous data is created by IoT connected devices. These data are stored in a big data system on a large scale. The big amount of data generated from IoT connected devices strongly depends on 3V factors or characteristics of big data that are volume (Data at rest), velocity (Data at motion) and variety (Data in many forms).
2. A Big data system is essentially a shared and distributed database, thus the tremendous quantity of data is filed in big data files in the storage system.
3. Advanced analytic tools like Hadoop, Spark, etc., are used for interpreting and examining the accumulated IoT big data.
4. The descriptions of examined data are inspected and generated for accurate and timely decision making.

Advantages of Data Analytics

1. Improved decision making
2. More effective marketing
3. Better customer service
4. More efficient operations

(f) Types of Data Analytics

The different types of data analytics that can be used and applied in the IoT investments to gain advantages are given below.

- **Streaming analytics:** Also known as event stream processing helps to analyze huge in motion datasets. In this type, urgent situations and immediate actions are detected by analyzing real data streams. Financial transactions, air fleet tracking, traffic analysis, etc., based IoT applications can use this method.
- **Spatial analytics:** The geographic patterns can be analyzed to determine the spatial relationship between the physical objects by this method. Location based IoT applications, such as smart parking applications use this form of data analytics.
- **Time series analytics:** In this method, time based data is analyzed to reveal associated trends and patterns. IoT applications, such as weather forecasting applications and health monitoring systems use this form of data analytics method.
- **Prescriptive analysis:** It is a combination of descriptive and predictive analysis. It is applied to understand the best steps of action that can be taken in a particular situation. Commercial IoT applications use this data analytics to gain better conclusions.
- **Descriptive analytics:** It is the most basic form of analytic insight that allows the user to describe and aggregate incoming IoT data. In descriptive analytics even calculations as simple as mean and standard deviation can be used to analyze the collected data. Description analytics might be used to answer the question, "What are the average pump temperature, flow rate, and RPM over a 30 minute time period?" in a connected factory use case.
- **Predictive analytics:** Predictive analytics is used to model future data and behaviors by analyzing historical data. Predictive analytics might be used to answer the question, "What is the estimated time to failure for a pump that is demonstrating a 20% increase in measured temperature?" in regression analysis such as linear regression. When identifying best in class predictive analytics capabilities on an IoT platform, enterprises should evaluate:
    - *On platform predictive analytic model building:* The ability of the platform to automatically or through programmatic interfaces generate a predictive model of the underlying platform obtained from the IoT data. Models such as linear or polynomial regressions are common, although more complex modeling choices are available in sophisticated platforms.
    - *On platform predictive analytic model operation:* The ability of the platform to utilize either a platform generated or platform integrated

data model (such as R or Python) to classify data or identify outliers through anomaly detection. Emphasis on the ability to manage models such as model versioning and updating as well as the ability to integrate a predictive model within a complex event processing (CEP) framework should be placed by the user.

- **Prescriptive analytics:** Prescriptive analytics are analyses that help enterprises to optimize the future direction to be taken. Some of the techniques used to complete prescriptive analytics are image processing, machine learning, and natural language processing. Prescriptive analytics might be used to answer the question, "To maximize pump uptime and minimize service intervals, what is the maximum allowed temperature increase for a pump before a preventative pump servicing must be scheduled?" When identifying best in class prescriptive analytics capabilities on an IoT platform, enterprises should evaluate:
  - On platform prescriptive analytic model capabilities: The ability of the platform to utilize either a platform generated or platform integrated data model, such as R or Python, to optimize a business outcome or relevant KPI. A prescriptive model should maximize or minimize a business-relevant KPI, such as time to delivery in route planning or equipment uptime for predictive maintenance.

Data Analytics helps enterprises to create business value by better understanding data. While there is no perfect IoT platform, some platforms are built with higher quality IoT analytics microservices than others. The best way to understand the capabilities of an IoT platform is by using them.

(g) Data Analytics Technology

Some of the modern data analytics technologies are given below.

- **Machine learning:** Artificial intelligence (AI) is the field of developing and using computer systems to complete tasks using simulated human intelligence. Machine learning (ML) is a subset of AI that is significant for data analytics and involves algorithms that can learn on their own. ML enables applications to take in data and analyze it to predict outcomes without someone explicitly programming the system to reach that conclusion. Initially, a small sample of data can be used to train a machine learning algorithm. The system will continue to learn as it gathers more data, becoming more accurate as time goes on.
- **Data management:** Before analyzing data, the procedures have to be in place for managing the flow of data in and out of the systems and keeping the data organized. The data should be of high quality and it should be collected in a central data management platform (DMP) where it is available for use when needed. Establishing a data

management program can help ensure that the organization is on the same page regarding how to organize and handle data.

- **Data mining:** The term data mining refers to the process of sorting through large amounts of data to identify patterns and discover relationships between data points. It enables to sift through large datasets and Figure out what's relevant. The information obtained can be used to conduct analyses and inform smart decisions. Data mining technologies allows the user to complete these tasks exceptionally quickly.
- **Predictive analytics:** Predictive analytics technology helps to analyze the historical data to predict future outcomes and the likelihood of various outcomes occurring. Statistical algorithms and machine learning are used in predictive analytics. More accurate predictions mean that the businesses can make better decisions moving forward and position themselves to succeed. It allows them to predict the future trends and stay ahead in competition by anticipating their customers' needs and concerns.

(h) AWS IoT Analytics for IoT devices

AWS IoT Analytics is one of the methods for IoT Data Analytics. AWS IoT Analytics is a fully managed service that makes it easy to run and operate sophisticated analytics on massive volumes of IoT data without having to worry about the cost and complexity to build an IoT analytics platform. It is the easiest way to run analytics on IoT data for IoT applications and machine learning and get insights to make better and more accurate decisions.

IoT data is highly unstructured. This makes it difficult to analyze using traditional analytics and business intelligence tools that are designed to process structured data. The fairly noisy physical processes such as temperature, motion, or sound are the common IoT data that are recorded by the devices. These data frequently have significant gaps, corrupted messages, and false readings that must be cleaned up before analysis can occur. Also, IoT data is often only meaningful with additional third party data inputs. For example, in the IoT application to help farmers determine when to water their crops, vineyard irrigation systems often provide moisture sensor data with rainfall data from the vineyard, allowing for more efficient water usage while maximizing harvest yield.

Each of the difficult steps that are required to analyze data from IoT devices automated using AWS IoT Analytics. It uses filters, transforms, and enriches IoT data before storing it in a time-slotted data store for analysis. Figure 7.16 gives the flow of working progress of AWS IoT data analytics.

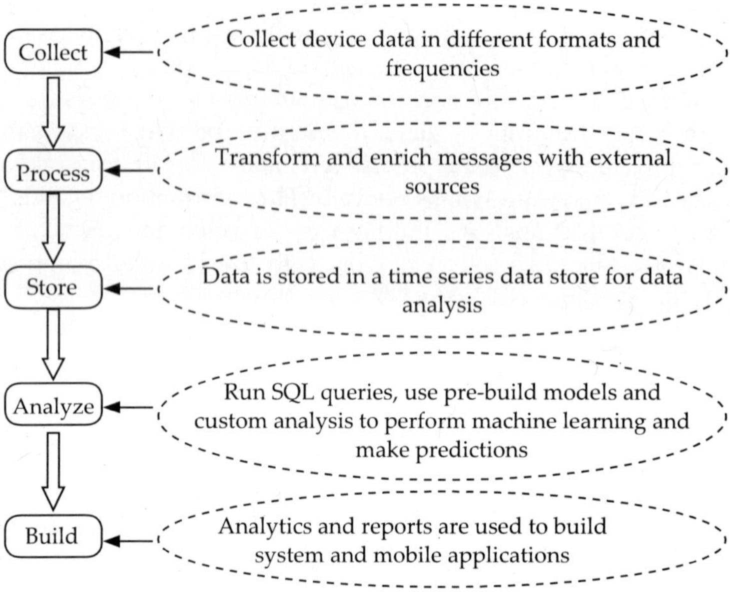

Fig. 7.16 Working progress of AWS IoT analytics

(i) Working Progress

- The working progress of AWS IoT analytics sets up the service to collect only the data needed from the devices, apply mathematical transforms to process the data, and enrich the data with information specific to the device such as device type and location before storing the processed data. Then analyze the stored data by running ad hoc or scheduled queries using the built-in SQL query engine, or perform more complex analytics and machine learning inference. Pre-built models in AWS IoT Analytics by including for common IoT use cases makes it easy to get started with machine learning.
- AWS IoT Analytics can also be executed using custom analysis packaged in a container. AWS IoT Analytics automates the execution of custom analysis created in Jupyter Notebook or own tools (such as Matlab, Octave, etc.) to be executed on schedule.
- AWS IoT Analytics is a fully managed service that operates on analysis and can scale automatically to support of up to petabytes of IoT data. AWS IoT Analytics can build fast, responsive IoT applications with data analytics obtained from millions of devices without managing hardware or infrastructure.

(j) AWS IoT Analytics Benefits

- Analytical workflows can be made operational.
- Prepare IoT data for analysis.
- Runs queries on IoT data easily.
- Tools for machine learning are available.
- Data storage is optimized for IoT.
- Pricing can be made with automated scaling with pay.

7.7 SUMMARY

This chapter discussed the required tools, hardware boards for developing IoT model. Then the developed model communicated with cloud for data access and the data to be analysed were illustrated.

REVIEW QUESTIONS

1. List the features of Python.
2. Write a program in Python to print Fibonacci numbers.
3. Write a program in Python to blink LED.
4. Write a program in Python to make buzzer.
5. Write a program in Python to make 5ms delay.
6. Compare the features of Arduino / Raspberry Pi.
7. What is the OS of Raspberry Pi?
8. Is it necessary for any OS to install Arduino.
9. Discuss about Contiki OS and Cooja simulator.
10. List the steps to interface IO devices with Raspberry Pi.
11. Write a program to display sensor data in Cloud platform.
12. List the Cloud servers.
13. Define data analytics in IoT.
14. Who is the first user of Cloud?
15. What is AWS?

8

BUILDING IoT ECOSYSTEM

8.1 INTRODUCTION

Hope all have gained a clear idea about IoT transmission along with its communication and security from the discussion of last chapter. In this chapter a summary of IoT ecosystem along with its reliability values are planned to discuss. The ecosystem (Figure 8.1) has three layers. The bottommost layer of the IoT ecosystem is the physical/perception layer which consists of various IoT physical devices (group of sensors/ actuators). The next layer is middle layer which is used to acquire sensed data from physical layer and then sent to the network via perfect transmission line. The third layer is an application layer which gets analyzed data from the middle layer, likewise the process goes on. A brief overview of all these processes is given as follows.

IoT devices are the basic building blocks of an IoT system. These devices are present in all the layers of IoT architecture. IoT devices can be classified as open-source or proprietary. Open-source IoT devices are those which are publicly available. Everything regarding open-source devices can be studied, modified and reproduced according to the requirement whereas proprietary IoT devices are licenced and details of these devices are not available. Mostly available IoT low end devices are sensors, actuators, Telos, Open mote, Wasp mote, Tmotesky, etc.

8.1.1 IoT Gateways

IoT gateway is a middle-end device. It includes Arduino yun, Netduino, Esp8266. It acts as an intermediate between various sensing networks and high-end IoT devices or to the Cloud platform through the Internet. The main purpose of these gateways is to manage the heterogeneity due to a variety of sensors data obtained from different sensors and sends data to the higher controller or platform. The data collected by the gateway should be filtered and processed. High-level processing of data is done based on the requirement of the applications. Gateways act as a major bridge between different layers of IoT architecture that lies in different networks. Gateways manage it and the terminal nodes attached to it and form a robust layer in the IoT ecosystem. In terms of data transfer gateways act as a proxy between the data centers or Cloud data source

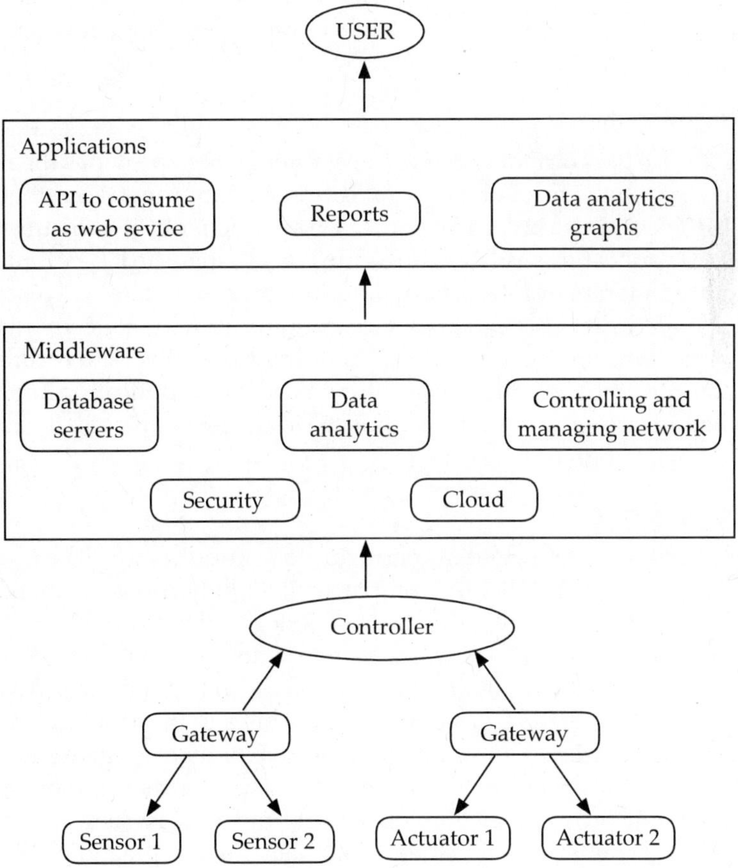

Fig. 8.1 IoT ecosystem

and the low-end IoT devices like sensors. Gateways sometimes act as a high-end device when the IoT ecosystem is large and more often acts as a middle-end device when the IoT ecosystem had a small number of low-end devices. Gateways have the ability to overcome failure and try to solve the communication gap between low-end devices and controllers.

8.1.2 IoT Communication

IoT devices are resource constrained. Due to resource constrained nature, there are many communication challenges like the requirement of low power communication, less memory-intensive routing protocol, deploy enhanced Wi-Fi, addressing and identification and high speed. If IoT devices connect through the Internet directly, it would require IP stack with high power and memory intensive. So, for making a perfect communication between IoT devices connection to be established using low power technologies like Wi-Fi, Zigbee, Bluetooth, NFC, low power technologies and WSN.

8.1.3 IoT Middleware

IoT middleware is a component of IoT ecosystem which provides the raw data to the user in the form of web application. The information gathered from IoT devices must be processed in a form such that it can support all types of applications related to the IoT ecosystem. IoT has limited functionality in terms of interpreting and integration of data. It presents a big challenge in terms of inter-device communication, collecting and accessing the data, and merging the data of multiple devices to form the flexible composition. IoT middleware for a particular application should be designed in such a way that it should be scalable, adaptable, flexible, secure and open source.

8.1.4 IoT Platforms

IoT platforms are the essential parts of IoT middleware. These are the specially designed software for particular applications to manage the complete system. They are the backbone of the IoT system. IoT platform connects devices, gateways, networks to the Cloud, server and applications. It helps the recognition of all devices. They provide a software development kit for all IoT system devices. This system enables virtual visualization of IoT devices from the perspective of managing, designing and enabling the IoT. They are like a bridge to the Cloud and the hardware IoT ecosystem. It provides a complete control of the ecosystem in terms of speed, cost, complexity, business and data flow are handled here. IoT platforms are the ones which enable the cross-device compatibility, automation of connected devices. The major tasks are to identify devices and to understand heterogeneous inputs of different devices over various protocols and keep the system updated.

8.1.5 Data Storage

Storing the large amount of data in various forms is one of the abilities of middleware. The demand of data traffic in IoT is high volume and thus requires new calibration and analytical techniques to support big data. Data mining techniques and its algorithms enable the complex computational processes of useful information for increasing raw data of IoT. By the way Cloud technology can manage and meet the complex data storage situation. In some cases transferring of data to the edge devices to reduce the data traffic in Cloud is also used. Fog computing and big data analytics are used for this purpose.

8.1.6 Security

Security is the degree of the special considerations to protect the individual's information from exposure to open environment. For this purpose mostly all IoT systems are provided with unique identification and the autonomous ability to communicate over all kinds of network bidirectionally. The privacy should not be compromised and present at all stages like in device, communication, storing, processing and its the applications. It is also possible to provide a security and privacy in middleware level. The network then broadcasts all the gathered information and events to the server for the logical and analytical process. Keeping the user's privacy and protection of data are one of the biggest challenges to be addressed in the IoT.

8.2 ENERGY EFFICIENCY AND RELIABILITY IN IoT

A sensor node uses its energy to carry out various functions such as data acquisition, communication and data processing. Acquisition of data depends on the type of monitoring activities used for the application. Communication of sensor data consumes more energy than other task. It covers in terms of emission and reception. Data processing is a technique in which an intermediate node is chosen to consolidate the sensor data streams from the source nodes to the sink node.

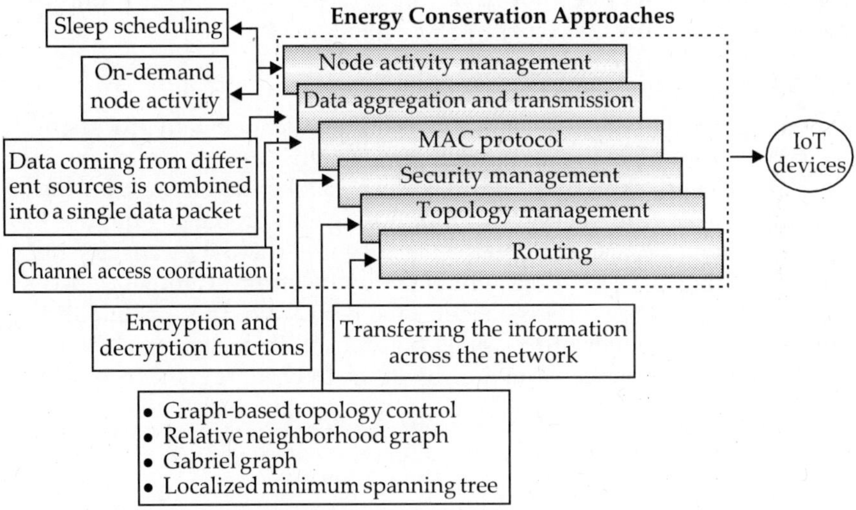

Fig. 8.2 Energy conservation approaches in IoT

(https://www.researchgate.net/Fig./Diagrammatic-representation-of-energy-conservation)

The term "energy efficiency" covers different aspects of a system in IoT. The most common aspects are energy per correctly received bit that is how much energy is spent to transport one bit of information from source to the destination, energy per reported event it is the energy spent to report one event, Delay/energy trade-off the notion of urgent events and speed in reporting such events and at least network lifetime time in which it is able to fulfill its task.

Currently, various researchers are working towards the energy conservation in IoT systems. The researchers found a few problems that may cause unnecessary energy consumption. The problems are idle listening, data collision and recovery, overhearing, reduction of protocol overhead, and traffic fluctuation.

In order to sort out from the above problems some researchers also found various ways as given in Figure 8.2 to get rid of the issues are discussed below.

8.2.1 Node Activity Management

There are two parts in node activity, namely, sleep scheduling and on-demand node activity. Sleep scheduling is the way to set the node to a sleeping mode and determine the time to wake up. This saves energy in idle time spans. Here, certain periods are determined in which the individual node is in sleeping mode. On-demand node activity is not scheduled, but the node is by default in an active state with a simple functionality. If a wake-up signal is broadcasted, the neighboring nodes within the area switch to active mode. After activation, the data transmission takes place. As the start-up signal does not have to be decoded, all the surrounding nodes are switched on.

8.2.2 Data Aggregation and Transmission Process

In general, the cost of data transmission is higher than the data processing. In order to minimize the cost number of transmission are done by clustering. Clusters shown in Figure 8.3 can reduce the amount of data, as the cluster heads are in-charge of monitoring and processing queries. It eases several energy dissipating effects caused by the data transmission. In this clustering process, data which is coming from different sources is combined into a single data packet. This helps to reduce redundancy. By the way, it is beneficial to aggregate data within clusters. Wireless transmission of data uses a large portion of the total energy. Incorporating power control into the transmission process could yield higher energy savings.

8.2.3 MAC Protocol

MAC protocol is considered a sublayer of the data link layer. It defines the rules to transmit the frame. If there are many nodes, the MAC protocol

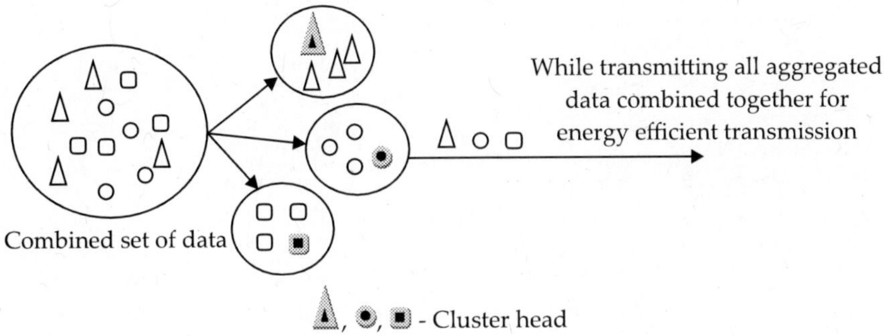

Fig. 8.3 Data aggregation method for energy harvesting

coordinates the channel access. Better design of MAC protocol is one of the ways to use energy efficiently. The important attribute of MAC protocol is energy efficiency. IEEE 802.15.4 defines two operation modes, i.e., non-beacon enabled mode and beacon enabled mode. The non-beacon enabled mode is always awake to receive a frame. The latter beacon enabled mode describes superframes where nodes are only awake during a small portion of a superframe. This causes increase in energy consumption and throughput. To achieve low power operation, several MAC protocols are proposed. The types of protocols for IoT are MQTT, XMPP, DDS and AMQP.

8.2.4 Security Management

In order to perform the encryption and decryption functions energy consumption has a significant impact on it. Requirement of security is necessary in each layer of IoT from the perception layer to application layer security is essential so the energy conservation in each layer. On behalf of security several steps has been taken by the researchers to sort out from the problem. Some of these lightweight cryptographic algorithms are suggested to be implemented in this layer to conserve energy.

8.2.5 Topology Management

The role of topology control is to reduce the node power consumption and extend the network lifetime. There are four different types of topologies to conserve power problem.

- **Graph based topology control:** This is done locally if information about distances between sensors and their relative position is available.
- **Relative neighborhood graph:** This is a straight line graph that connects two points. The triangulation of a point is a maximal of non-intersecting line segments with vertices in the point.

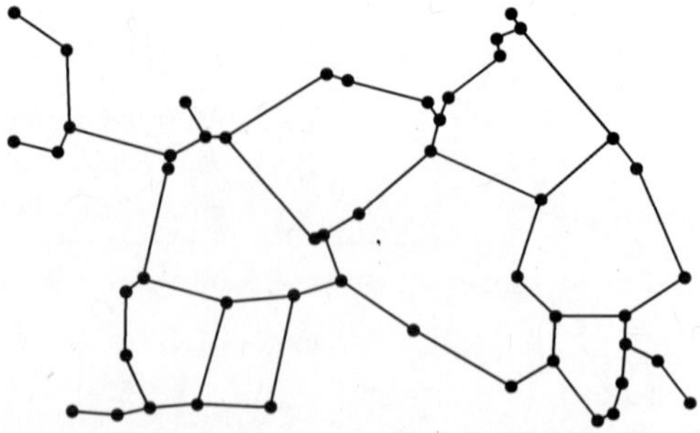

Fig. 8.4 Model of relative neighborhood graph.

- **Gabriel graph:** In this graph every sensor in the network knows its neighbors and their locations. Each sensor determines its logical neighbor by computing the closed discs of diameters.

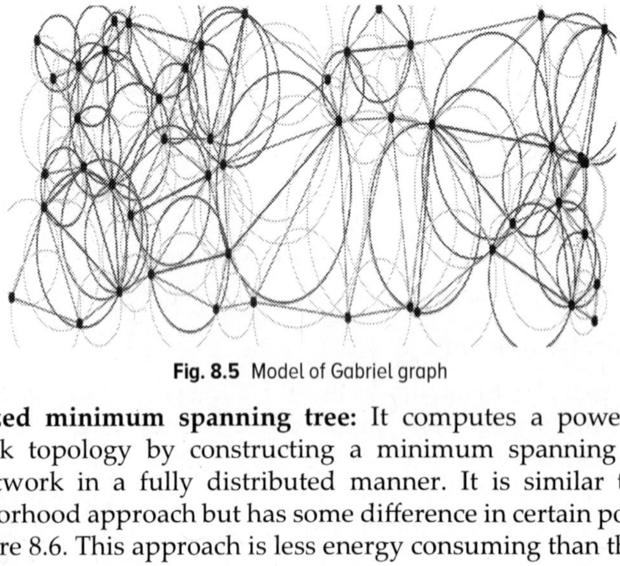

Fig. 8.5 Model of Gabriel graph

- **Localized minimum spanning tree:** It computes a power reduced network topology by constructing a minimum spanning tree over the network in a fully distributed manner. It is similar to relative neighborhood approach but has some difference in certain points given in Figure 8.6. This approach is less energy consuming than the original network.

Relative neighborhood graph

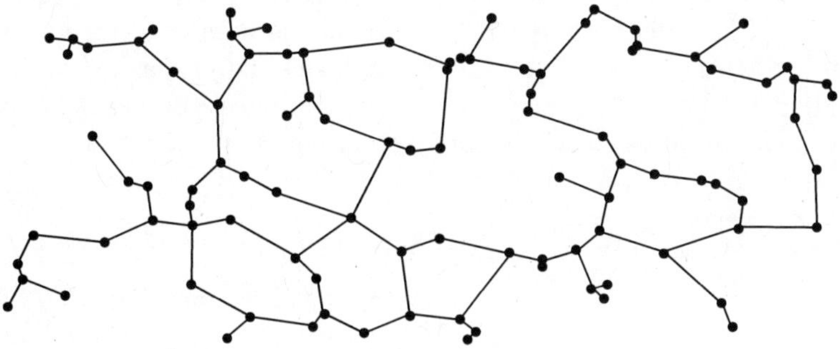

Local minimum spanning tree

Fig. 8.6 Model of local minimum spanning tree with its difference from relative neighborhood graph

8.3 WIRELESS RF CHARGING

Nowadays, the more fascinating technology is wireless charging of mobile phones. Many of the cars (more than 50 car models) as shown in Figure 8.7 have been coming up with this feature. In this method one can charge our mobile phones by simply keeping into the trays. There is no need of power cables to connect the mobile to plug point. The main concept behind this working is the charging tray emits the radio waves which are picked up by the mobile phones and then convert this acquired radio waves into electrical signal in terms of charging. The detailed explanation and working of RF energy harvesting given as follows:

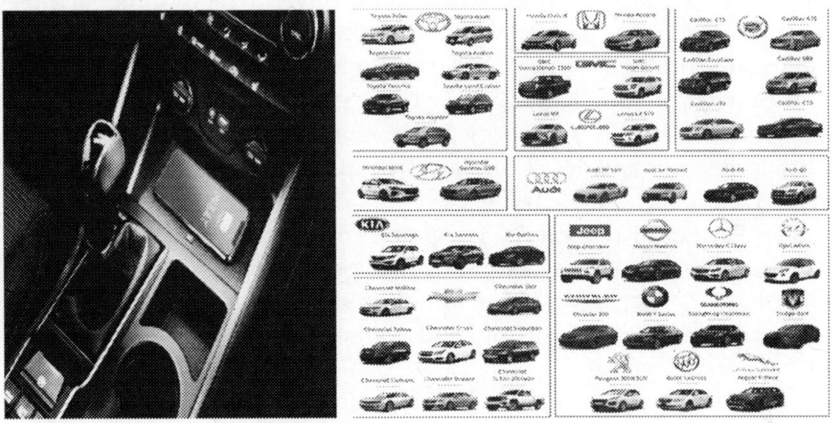

Fig. 8.7 Model of Wireless charging in car

8.3.1 RF Wireless Charging (RF energy harvesting)

In RF wireless charging as in Figure 8.8 the in-built antenna in the electronic device picks up low level radio frequency waves from external sources and converts the energy of the waves to direct current (DC) voltage. The DC voltage, the antenna harvests over the air, is then used to power an electronic device or recharge the device batteries. This type of charging is well-suited for little DC power consumption devices, powering sensors and other electronic devices that run on very small power.

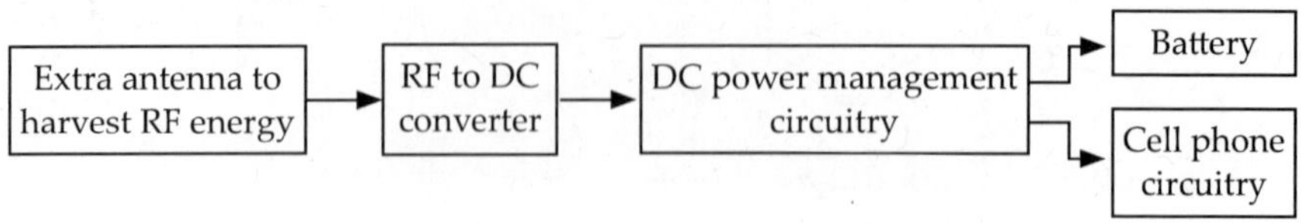

Fig. 8.8 Work flow of RF charging

Radio frequency (RF) wireless charging technology is unique because it uses electronic waves or radio frequencies rather than magnetic fields to charge a device. An RF transmitter transmits RF waves and a receiver embedded within a device picks them up. The RF waves are then converted into electricity which powers the device. RF waves can power devices at a distance ranging from a few centimeters to a few meters.

The concept of wireless power transmission was originally proposed by Nikola Tesla, a Serbian-American electrical engineer and inventor during the 19th century. He developed and promoted alternating current (AC) recently his efforts to make vision a reality have begun with approaches that use tightly coupled induction or loosely coupled magnetic fields resonance to transfer energy. Induction and resonance, which harness the power of, require the transmitter and receiver to be in direct or close contact, respectively. Smart phone charging pads and tablet charging stands are examples of commercial uses for this type of wireless charging.

In contrast, RF wireless charging harnesses the power of waves in the radio spectrum to transfer energy and creates a charging zone that's very similar to a Wi-Fi hotspot. There are two main categories of RF wireless charging: Far Field and Near Field

1. In Far-field RF wireless charging transmission takes place in open space that is as long as the device located in a particular area the device gets charged (the working is similar to a Wi-Fi router). The main limitation of this method is that a great amount of energy is lost in free space.
2. In Near Field RF wireless charging allows any 3D object to become a charging station (transmitter). It provides a greater spatial freedom and a true "drop and charge" experience. The charge loss in free space is also reduced.

Wireless charging is finding applications in the healthcare, automotive and manufacturing industries because it offers the promise of increased mobility and advances that could allow tiny IoT devices to get power from many feet away from a charger.

Key Benefits

RF offers many key benefits and are listed as follows.

(a) **Spatial freedom:** RF wireless charging technology provides spatial freedom ranging up to 15 feet.

(b) **Seamless integration:** Unlike induction and other forms of wireless charging, RF wireless charging is not restricted by measurements by the way it does not have any physical size restrictions of devices.

(c) **Multi-device charging:** With RF, several devices can be charged at the same time, through a defined charging zone, using a single transmitter. Extended charging zones and different power requirements can also be accommodated. This gives consumers a common charging source for the multiple devices in their connected lives.

(d) **No installation:** It does not require any proper installation processes. It just requires a charging tray alone. There is no need for precise placement as there is with plugs or inductive charging pads. This means that no matter how you place your device within a certain area, it will charge regardless of how the device is placed. It is more convenient for electronic devices and it makes them smarter too.

8.3.2 Improving RF Energy Transfer Efficiency

A lot of recent research works have focused on investigating the RF-DC conversion efficiency of rectifier circuits due to the increased number of applications which are related to the IoT. The RF-DC conversion efficiency of a rectifier using modulated signals is to maximize the power transfer efficiency based on the used signal properties and output load. Recent works have already shown that the RF-DC conversion efficiency of a rectifier is strongly affected by the underlying signal PAPR (peak-to-average power ratio). The roll-off factor has a strong effect on PAPR, which influences the RF-DC conversion efficiency.

The optimum load corresponding to a maximum RF-DC conversion efficiency increases with increasing PAPR of the input signal. As a result, signals with lower PAPR tend to lead higher efficiency at lower load values while for large load values the efficiency obtained with signals having a high PAPR is better. It is also observed that while PAPR is a good indicator of the RF-DC efficiency performance, the complete CCDF (complementary cumulative distribution function) curve, or other statistical measures such as the instantaneous power variance is necessary for an accurate evaluation of its behavior. Finally, the RF-DC conversion efficiency depends on the modulation rate through the output filter of the rectifier formed by the output capacitor and load value. It is understood that depending on the CCDF and PAPR properties of the applied signal a different optimum load exists which leads to a maximum efficiency.

8.4 CLOUD BASED IoT STACKS

8.4.1 The IoT technology Stack

The IoT technology stack is a range of technologies, standards and applications. It starts with the device hardware like sensors that are connected to the Internet till the most complex applications that are developed to use these connected things. The flow of IoT communication is given in Figure 8.9.

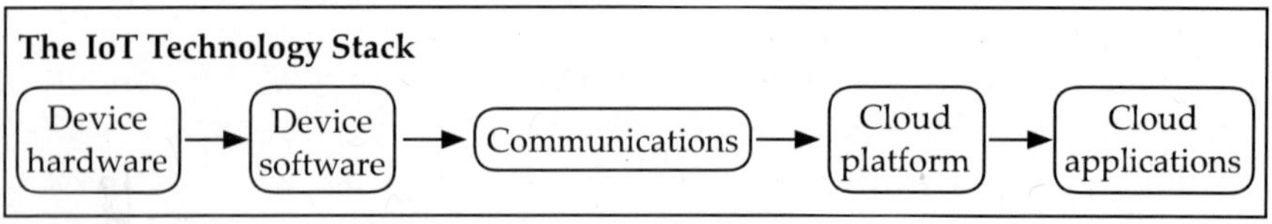

Fig. 8.9 IoT Technology stack

Image Credit: The IoT Magazine

The five layers in a IoT technology stack are:

- Device hardware
- Device software
- Communications
- Cloud platform
- Cloud applications.

1. ***Device Hardware:*** The device hardware is the 'thing' that is to be connected to the Internet. The main aim of the device is to collect data. Based upon the amount of data to be collected, the device might be simple like a single smart sensor for simple data collection and for more complex data collection; an industrial computer that houses many sensors, a powerful processor, local storage, a gateway, etc., can be used.

   At this level of the stack, it is important to understand some implications of cost, size, ease of deployment, reliability, useful lifetime, etc. For example, for small devices a system on a chip (SoC) is sufficient while for more demanding solutions, an embedded computer like Artik module, Raspberri Pi, Arduino or BeagleBone board is needed. For really serious computing needs, advanced industrial computers like compact RIO or PXI are needed.

2. *Device Software:* The conversion of device hardware into a 'smart device' is done by the device software. This is the part of the IoT technology stack that enables the concept of "software-defined hardware", meaning that a particular hardware device can serve multiple applications depending on the embedded software it is running on.

   The communication to the Cloud or to other local device is implemented using the device software. Real-time analytics, data acquisition from the device's sensors, and control can be done in this layer. This part of the IoT technology stack is vital because it serves as a bridge between the real world (hardware) and the Cloud applications.

   The device software layer can be divided into two categories:

   - Operating system
   - Applications

   (a) **Operating System:** The type of operating system (OS) is determined by the complexity of the application for which it is to be used. Some of the key considerations to determine the operating system include whether the application needs a real-time OS, the type of I/O support needed, and whether support for the full TCP/IP stack is needed. Common examples of embedded OS include Linux, Brillo (scaled-down Android), Windows Embedded, and Vx Works.

   (b) **Applications of OS:** These are the applications that run on top of the OS and provide the functionality that is specific to the IoT solution. The possibilities are endless for applications of OS. Data acquisition and streaming to the Cloud, analytics, local control, etc. are a few.

3. *Communications:* In this part of the IoT technology stack, need to define the network communication platforms that will connect the sensors in the product hardware to the Cloud and then to the application.

   Communications refer to all the different ways a device can use to exchange information with the rest of the world. This includes both the physical networks and the protocols. Selecting the right communication mechanism is a key part of constructing the IoT stack. It will determine not only how to get data in and out from the Cloud but also on how to communicate with other third party devices in the same environment. The communication layer has an impact on the overall topology of the system.

   Communications are not purely technical decisions. They are made by considering the impact to cost, deployment, and technical complexity of the solution. There are many different types of connectivity that we can use, including Bluetooth, radio, Wi-Fi, cellular, satellite.

Some of the examples of communication protocols include:

- Infrastructure (ex: 6LowPAN, IPV4/IPV6, RPL)
- Identification (ex: EPC, μCode, IPV6, URIs)
- Transmission (ex: Wi-Fi, Bluetooth, LPWAN)
- Discovery (ex: Physical Web, mDNS, DNS-SD)
- Data protocols (ex: MQTT, CoAP, AMQP, Websocket, Node)
- Device management (ex: TR-069, OMA-DM)
- Semantic (ex: JSON-LD, Web Thing Model)
- Multi-layer Frameworks (ex: Alljoyn, IoTivity, Weave, Homekit)

4. *Cloud Platforms:* The Cloud platform is the backbone of any IoT solution. The infrastructure will serve as the platform for these key areas:
   (a) **Data Collection and Management:** The smart devices will stream information to the Cloud. Based on the requirements of the IoT solution, the type and amount of data that can be collected on a daily, monthly, and yearly basis can be determined.
   One of the challenges of IoT applications is that they can generate an enormous amount of data. The scalability parameters for data management have to be determined from the very beginning for the right data management solution.
   (b) **Analytics:** Analytics is one of the key components of any IoT solution. It is the process of breaking down data, finding patterns, performing forecasts, integrating machine learning, etc. It is the ability to find insights from existing data.
   (c) **Cloud APIs:** The IoT is all about connecting devices and sharing data. This is usually done by exposing APIs at either the Cloud level or the device level. Cloud APIs allow customers and partners to either interact with devices or to exchange data. Additionally, Cloud analytics needs to process incoming data in real time, to detect the trends and be able to make predictions.
5. *Cloud Applications:* IoT applications are only limited by creativity. They can be of any form like: Hardware, software, Cloud, etc. This is the part of the stack that includes the end-user applications which the customer will see and interact with. These applications will most likely be web based, and depending on your the needs, separate apps for desktop, mobile, and even wearable can be developed.
   Applications can be divided into customer facing and internal apps. Customer facing applications usually get the most attention, but in

the case of IoT, internal applications are equally important. These include applications to remotely provision and troubleshoot devices, monitor the health of your device fleet, report on performance and predictive maintenance, etc.

Major applications are listed below.

Smart Home, Wearable, Smart city, Smart Grids, Industrial Internet, Connected cars, Smart supply chain, Connected Health, Smart retail, Smart farming.

8.4.2 Data Management

It is the process of getting the overall available data from an IoT application and refining it down to important information. Different devices from different applications send large volumes and a large variety of information. All this IoT data can be managed by means of developing and executing architectures, policies, practices and procedures that can meet the full lifecycle data needs.

The edge devices between the main database and end node help to secure the most valuable data and reduce the bandwidth cost by preventing more data to the Cloud. It provides a great performance, ownership over data and lower maintenance cost. These high-performance systems can not only collect data in real time but also organize and provide data to other systems.

The data is usually processed near the data source or at the edge of the network in edge computing. Sometimes in case of lesser data process takes place in a centralized data storage location in a typical Cloud environment. By doing processing in local area the IoT saves storage space for data and processes information in the Cloud meanwhile it helps to maintain the security challenges.

Manage data on the edge or in Cloud decisions can be taken based on scalability and agility, security and usability of data. After this analysis the decision can be met with processing at edge computing, data governance policies and metadata management.

In edge devices the end-users can access to monitor the flow of data and decide how various systems in demonstration and devices are running. It is a web based activity. The end users have a provision to get notified by alarms. A large amount of data can be represented in the form of a graph for any desired range of time, and each point on the graph represents a record that can be found by searching the database, which stores a large quantity of data. Some of the data management challenges are:

- If number of IoT devices increases it increases the challenges for real-time processing and analysis to reduce time for storage. Space has to be optimised for metadata like user IDs and passwords to ensure enough space for new information.

- Data management frequently requires an adaptive maintenance, predictive repair, security monitoring and process optimisation rely on real-time data. Selecting the right tools is a challenge because integration between different sensors should be proven and compatibilities confirmed.
- Interoperability, scalability, security and standards are offered by software technologies some of the important factors to be kept in mind while building IoT products. It is important to protect data from unauthorised access and tampering. Organisations need to be compliant with national rules and regulations on securing data.
- Likewise, quality of IoT device data has to be checked frequently to enhance the reliability of the device.

8.5 SUMMARY

The IoT ecosystem was discussed with all possibilities of system blocks, energy efficiency with reliability of IoT network and IoT slack management.

REVIEW QUESTIONS

1. What is ecosystem?
2. Differentiate between gateway and router.
3. List the energy conservation approaches.
4. What is meant by middleware in IoT?
5. Define data aggregation and transmission.
6. In which layer MAC protocol is used?
7. What is the need of security in IoT?
8. Describe node activity management.
9. Explain the working principle of RF wireless charging.
10. Write some example of wireless charging devices.
11. Define data management in IoT.
12. What is IoT stack?

9

APPLICATIONS AND CASE STUDIES

9.1 INTRODUCTION

Various applications of IoT models are developed and demonstrated with required tools and board.

9.2 HOME AUTOMATION

In this modern world everyone wants to control their home gadgets by using their own electronic gadgets. As far as from the earlier chapters we have understood that IoT is one of the emerging automation technologies. It allows us to control hardware devices via Internet. Some of the examples are lighting control, lawn/gardening management, improved home safety and security, air and water quality monitoring, natural language based voice assistants, better infotainment delivery, AI-driven digital experiences, smart switches, smart locks and smart energy meters. Home automation process depends on three important components for its seamless communication. It can be explained briefly as follows.

9.2.1 Components of Home Automation

The IoT based home automation is remotely controlled over the Cloud. In general, wireless home automation works by deploying the low power equipment to send and receive information and to control devices remotely. In this system the electrical and electronic appliances such as fan, lights, outdoor lights, fire alarm, kitchen timer, etc., can be controlled using our hand-held devices. There are three important components present in the home automation to convert conventional devices into smart devices (Figure 9.1).

Fig. 9.1 Components of home automation

Each of the above parts is equally important to build a comfortable smart home experience. It can be described as follows.

Hardware

The perfect hardware enables the ability to develop your own IoT prototype based on your application needs and iteratively responds to technology with ease of accessibility. Hardware part includes sensors and gateways.

Sensors: There are probably thousands of such sensors. In that most usable sensors are: temperature sensors, lux sensors, water level sensors, air composition sensors, video cameras for surveillance, voice/sound sensors, pressure sensors, humidity sensors, accelerometers, infrared sensors, vibration sensors and ultrasonic sensors.

Gateway: For developing a home automation product, often a stand-alone product sending data to a server is not enough. Due to battery and protocol limitations, the data from a sensor or sensors present in a home has been routed through an IoT gateway. To select the perfect gateway for IoT home automation we have to consider some of these factors:

- Supported communication protocols
- Real-time capabilities
- MQTT, CoAP, and HTTPS support
- Security and configuration
- Modularity.

Communication Protocols

It is one of the most important parts of a home automation product which is used to communicate between gateways, servers, and sensors. A protocol selected with the right testing and careful consideration helps to avoid performance bottlenecks. By the way, it would restrict the technology and device integration capabilities with sensors and IoT gateways. Another important consideration is the firmware (operating system) that resides in hardware which manages data transfer, firmware updates, and performing other critical operations to make things talk. Some of the communication protocols used for home automations are discussed.

When looking at the major home automation protocols, Bluetooth low energy, Z-wave, Zigbee, and thread are commonly preferred protocols and the selection of that suitable protocol can be narrowed down by the following factors:

- Ability to perform identity verification
- Quality of sensor networks
- Data transfer rate
- Security level
- Required network topology
- Density of objects around
- Effective distance coverage.

Software

Based on the user's need various apps can be created at the application side to simplify the whole operation in a hand-held device and maintain the Cloud database for storing and retrieving information. Cloud services are online services that provide high level APIs used to dereference various low level details of underlying network infrastructure like physical computing resources, location, data partitioning, scaling, security, backup, etc.

Let's have a detailed look into some of applications in the home automation.

9.2.2 Smart Energy Meter using Arduino

Wireless energy meter is mainly designed to reduce the time and labor consuming problems in normal energy meter.

Working of Conventional energy meter

The conventional energy meter (Figure 9.2) is used to measure the amount of electricity consumed by a residential/commercial building based on the amount of voltage consumed in certain time period. Initially, the reading was noted manually by a person from electricity board and then based on the reading the amount was calculated and noted to the user's card.

Fig. 9.2 Working of conventional energy meter

In smart energy meter the reading from the energy meter is transmitted to the user's mobile or mail via Internet with online payment or bill sharing. To implement this facility, the following components are required.

Hardware

1. Single-phase energy meter
2. Arduino MEGA2560 or UNO
3. GSM module (SIM900A)
4. LCD display
5. Jumper wires

Software used

Arduino IDE

Working

The smart energy meter (Figure 9.3) is designed using Arduino. Arduino takes the pulse values from the energy meter, and then calculates units and displays the reading on the LCD. The readings are stored in-built EEPROM so that in case of any power failure it continues reading. The reading of the energy meter is sent to the cell phone of the user by a message through GSM modem. In this work relay and relay drive are used so that if Arduino fails or stop working, it automatically discontinues power supply to user and we get the information about it. This smart meter is powered by external AC power supply from that it takes the AC power and converts it into DC power then it is fed to Arduino and GSM modem.

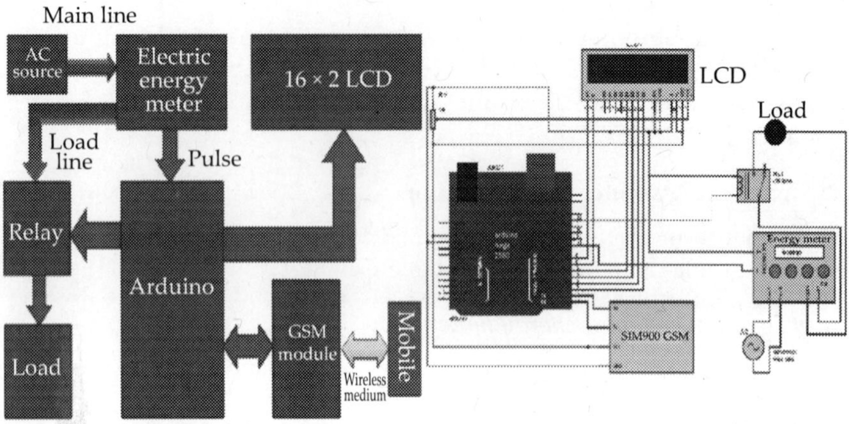

Fig. 9.3 Block and pin connection diagram of smart energy meter

(https://www.instructables.com/GSM-Based-Energy-Meter-Billing-Using-Arduino-Mega-/)

9.2.3 Smart Lock

In ordinary locks we use keys to open and close doors and can use number lock system to open the door. In smart lock system a perfect lock system was provided with utmost security. In case of any theft problem it is possible to send the information automatically to the nearby police station or neighbors. In this system by using Arduino just by a single login can open and close the doors with utmost security. The required components to implement are discussed as follows.

Hardware

1. Arduino UNO R3
2. ESP8266 with PCB antenna
3. Darlington high power transistor
4. Lock style solenoid

Connection

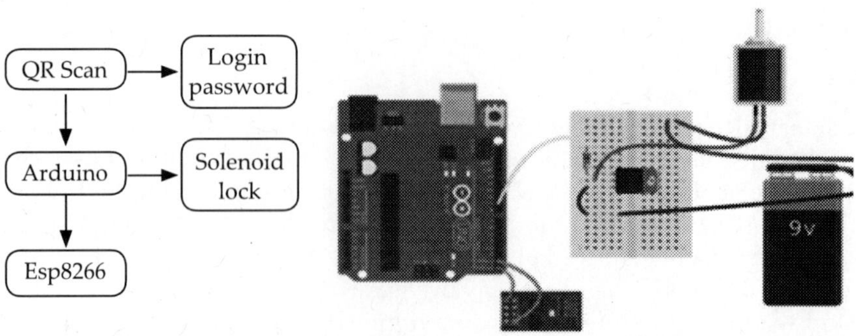

Fig. 9.4 Block and circuit diagram of a smart lock system

(https://create.Arduino.cc/projecthub/electropeak/smart-door-lock-using-Wi-Fi-login-page-by-Arduino-esp8266-e13825)

Working

Smart lock door system (Figure 9.4), initially, a login section for a door can be done by using Wi-Fi and solenoid lock. We don't use keypad or complicated mechanical elements for this system. We just stick a QR code (can be created online using some readily available online tools) to the door and then, allowed people can scan it to see the login page and enter their own confidential password. After typing the password, the solenoid lock will be activated either to open or close the door. This work uses an Arduino board and a driver for the solenoid and an ESP8266 to make a connection to local Wi-Fi. In case of any repeated wrong passwords will activate auto lock and forgotten passwords its possible to create some remembrance features in it.

9.2.4 Smart Lighting using Arduino and PIR Sensors

In normal lighting system a person has to go on and off the light and in some places the light glows unnecessarily, in order to avoid this problem the smart lighting systems helps a lot. In this smart lighting system uses Arduino and PIR sensor, the lights in the room will automatically turn ON and OFF by detecting the presence of a human, leads to power conservation and time consumption.

Hardware

- Arduino UNO
- PIR Sensor
- 5 V Relay Module (Relay Board)
- LED
- 100 Ω Resistor (1/4 watt)
- Connecting Wires
- Breadboard
- Power Supply

If you do not have a Relay Module, use the following components:

- 5 V Relay
- 2N2222 (or BC547) NPN transistor
- 1N4007 PN junction diode
- 1 KΩ Resistor (1/4 watt)

Software

- Arduino IDE

Connection:

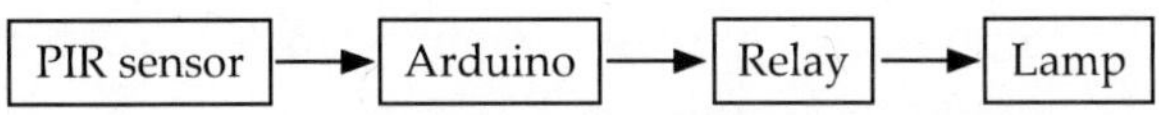

Fig. 9.5 (a) Block diagram

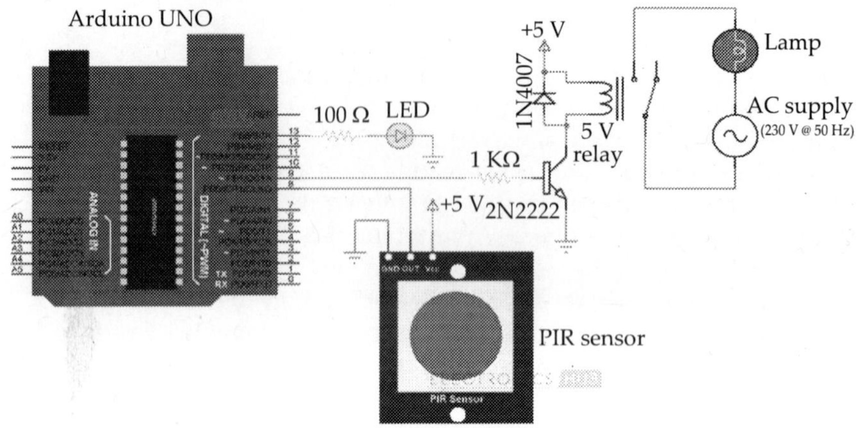

Fig. 9.5 (b) Circuit diagram of smart lighting system

(https://www.electronicshub.org/automatic-room-lights-using-Arduino-pir-sensor/)

Working

The block diagram of smart lighting is given in Figure 9.5. When there is no human movement, the PIR sensor doesn't detect any person and its OUT pin stays LOW. As the person enters the room, the change in infrared radiation in the room is detected by the PIR sensor. As a result, the output of the PIR sensor becomes HIGH. Since the Data OUT of the PIR sensor is connected to Digital Pin 8 of Arduino as shown in Figure 9.5 b, whenever it becomes HIGH, Arduino will activate the relay by making the relay pin LOW (as the relay module is an active LOW module). This will turn the light ON. The light stays turned ON as long as there is movement in front of the sensor. If the person takes a nap or leaves the room, the IR radiation becomes stable (there will be no change) and hence, the Data OUT of the PIR sensor will become LOW. This, in turn, will make the Arduino to turn OFF the relay (make the relay pin HIGH) and the room light will be turned OFF.

9.3 SMART CITY

Smart city is an area that uses different types of electronic methods and sensors to collect data. Information gained from the sensor is used to manage asset, resource and service efficiently. Many governments around the globe are racing to infuse technology into just about every aspect of its city's operation. Smart city involves development in various fields as given in Figure 9.6 are evolved like public transportation, IT connectivity, water, power supply, sanitation, solid management, efficient urban mobility, e-governance, citizen participation.

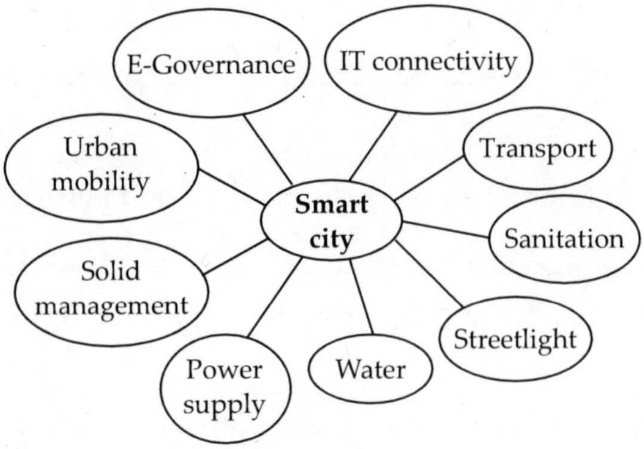

Fig. 9.6 Smart city

Objective of smart city

- Quality of life
- Economic competitiveness
- Sustainability
- Develop sustainable and innovative funding streams
- Empowered citizen

9.3.1 Smart and Safety Transportation

Transportation is an essential need of all human beings. Humans are always in the mode of hiring something for satisfying their needs. They travel for performing their day-to-day activities like going to their workplaces, schools, colleges, hospitals, entertainment places, etc. Humans usually prefer walking for shorter distance and travel by vehicle for longer distance. Walking or riding, safety is the most important thing while walking they may have to cross traffic signals and while riding they have to pass traffic signal in both these cases traffic signal plays a crucial role so that the first thing taken in this proposal is to manage traffic signal in an efficient and smart way to get rid of traffic congestion. Secondly, for safety transportation, drivers attention is also need to be considered monitored frequently and alerting him always. In addition to this having daily travelers in mind air pollution monitoring and alerting people to beware of exploiting and inhaling polluted air.

Objectives

- To handle heavier traffic in a smart way by extending green time.
- To provide congestion-free route for emergency vehicles.
- To provide expressway monitoring and advisory system to avoid traffic congestion due to some incident.

- To provide extended green man time for person with disabilities/aged person.
- To monitor vehicles speed limit and gas leakages frequently for safer driving.
- To monitor drivers attention frequently.
- To monitor air pollution to aware both public and person on board.

Methodologies/ Implementation Strategy

The method used here is mainly based IoT and some of the models are given as follows.

9.3.1.1 IoT Traffic Signal Monitoring and Controller System

Here IoT based automated traffic signal monitoring as well as controller system that automates complete traffic signaling system automation and also allows for manual override over Internet was proposed. The system given in Figure 9.7 uses Arduino based circuit system to monitor traffic signal densities and transmits this data online over Internet to the controllers. Use IoTGecko in order to develop the online GUI based system to monitor the traffic densities. The system shows current densities to help monitor traffic conditions on roads. Also the system provides an option to the controllers to override any signal and make it green in case of any ambulance or important vehicle to pass through while keeping other signals red. This puts forth a traffic signal monitoring and controller system that can be operated remotely over the Internet from anywhere with manual override ability.

Hardware Specifications

- ATmega328P AVR MC
- IR transmitter receiver

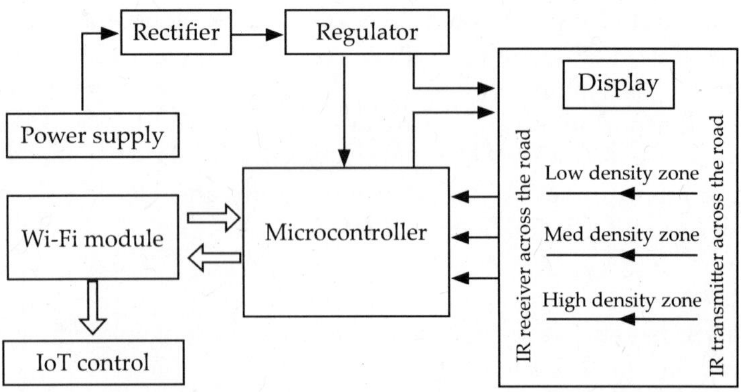

Fig. 9.7 Block diagram of IoT based traffic signal monitoring and controlling

- ESP8266 Wi-Fi module
- LCDs
- Crystal oscillator
- Resistors
- Capacitors
- Transistors
- Cables and connectors
- Diodes
- PCB
- Transformer/adapter
- Push button

Software Specifications

- Arduino compiler
- MC programming language: C

9.3.1.2 Vehicles Speed Limit Checker

In this the vehicles those are not following speed limits are identified by means of GPS inbuilt circuit in the vehicle and the vehicles number are captured by means of surveillance camera and passes this information to the traffic controller. Likewise, vehicles which are violating the traffic rules by crossing red signals are duly monitored.

EMAS (Expressway Monitoring and Advisory System): This advisory system is mainly used to monitor and control the various incidents taking place on highways or in some common place. The recovery team will come to the place at any time. In case of any major incident the information is transmitted to all junctions and reroute the vehicle as much as possible.

9.3.1.3 IoT Gas Pipe Leakage Detector Insect Robot

Gas pipes play an important role for cities, industries and thus in growing economies. So gas leakages lead to losses as well as are a threat because they can also lead to fire accidents. Placing sensors at each section of the pipe is costly. Figure 9.8 proposes an innovative robot that clings on to the outer surface of the gas pipe and moves with the pipe to check for leakages. The robot consists of a gas sensor that is used to detect gas leakage. As the robot keeps moving along the metal pipe it keeps monitoring for any gas leakage, On detection it uses an interface GPS sensor to transmit location of the leakage detected over to the IoT login system, Here use IoTgecko to receive and display the gas leakage alert and location over IoT. Thus, we have a fully automated insect-like robot that moves with the gas pipe and detects gas leakage instantly at a low budget.

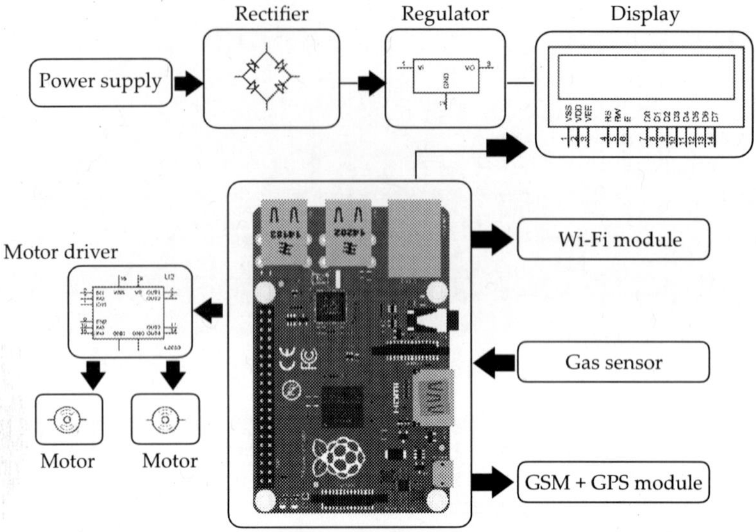

Fig. 9.8 IoT Gas pipe leakage detector insect robot

Hardware Specifications

- Raspberry Pi 3
- ESP8266 Wi-Fi module
- MQ-6 LPG CNG gas sensor
- SIM 800 GSM module
- Neo6mv2 GPS module
- Buzzer
- DC motor
- Motor driver IC
- Voltage regulator IC
- IC socket
- Resistors
- Capacitors
- Transistors
- Cables and connectors
- Diodes
- PCB
- LEDs
- Push button

Software Specifications

- Python compiler
- Programming language: Python

9.3.1.4 IoT Air and Sound Pollution Monitoring System

Air and sound pollution is a growing issue. It is necessary to monitor air quality and keep it under control for a better future and healthy living. Here, in Figure 9.9 proposes an air quality as well as sound pollution monitoring system that allows us to monitor and check live air quality as well as sound pollution in particular areas through IoT. The system uses air sensors to sense presence of harmful gases/compounds in the air and constantly transmit this data to microcontroller. Also these system keeps measuring sound level and reports it to the online server over IoT. The sensors interact with microcontroller which processes this data and transmits it over Internet. This allows authorities to monitor air pollution in different areas and take action against it. Also authorities can keep a watch on the noise pollution near schools, hospitals and no-honking areas, and if system detects air quality and noise issues it alerts authorities so that they can take measures to control it.

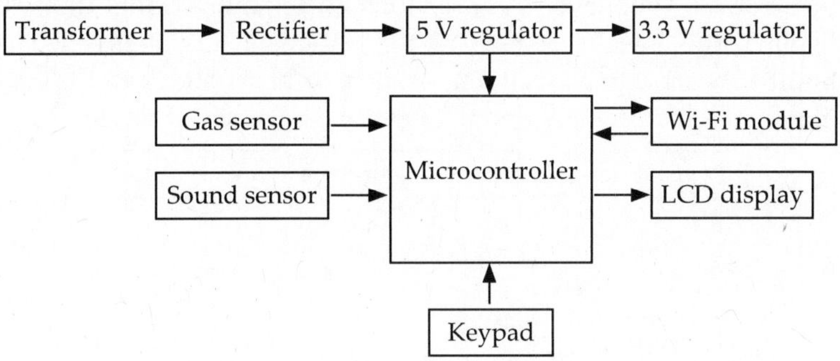

Fig. 9.9 IoT air and sound pollution monitoring system

Hardware Specifications

- ATmega328P AVR MC
- MQ 135 Sensor
- Mic Sensor
- ESP8266 Wi-Fi module
- LCDs
- Crystal oscillator
- Resistors
- Capacitors
- Transistors
- Cables and connectors
- Diodes
- PCB
- LEDs
- Transformer/adapter
- Push button

Software Specifications

- Arduino compiler
- MC programming language: C

9.3.1.5 IoT Heart Attack Detection and Heart Rate Monitor

Nowadays an increased number of heart diseases including increased risk of heart attacks are identified. Our proposed system in Figure 9.10 uses sensors that allow detecting heart rate of a person using heartbeat sensing even if the person is at home. The sensor is then interfaced to a microcontroller that allows checking heart rate readings and transmitting them over Internet. The user may set the high as well as low levels of heart beat limit. After setting these limits, the system starts monitoring and as soon as patient heart beat goes above a certain limit, the system sends an alert to the controller which then transmits this over the Internet and alerts the doctors as well as concerned users. Also the system alerts for lower heartbeats. Whenever the user logs on for monitoring, the system also displays the live heart rate of the patient. Thus, concerned ones may monitor heart rate as well get an alert of heart attack to the patient immediately from anywhere and the person can be saved on time.

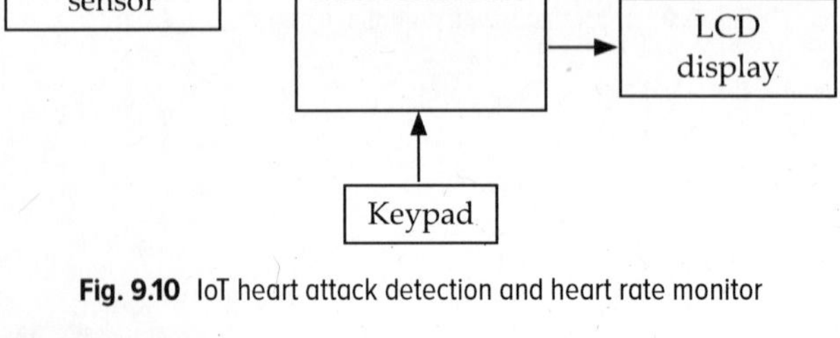

Fig. 9.10 IoT heart attack detection and heart rate monitor

Hardware Specifications

- ATmega328P AVR MC
- Heartbeat sensor
- ESP8266 Wi-Fi module
- LCDs
- Crystal oscillator
- Resistors
- Capacitors
- Transistors

- Cables and connectors
- PCB
- LEDs
- Transformer/adapter
- Push button

Software Specifications

- Arduino compiler
- MC programming language: C
- IoTGecko

In India, more than 1,40,000 people died in 2016 due to road accidents. USA and China like countries have the numbers are even worse. Many lives could have been saved if the passengers used seat belts and many accidents could have been prevented if the drivers were a bit more alert while driving. Our innovation offers a solution to save driver's life. The supporting system to implement are discussed further.

- **Intelligent Ignition System (IIS):** Seat belts are a primary form of safety in any four-wheeler. But people seem to ignore this form of safety and hence are putting their and other's lives in danger. 'IIS' allows the ignition of the vehicle, only if the seat belts are on. A magnetic encoder is added to the existing 'seat belt retracting system', along with a tiny magnet. The encoder records the rotation of the retracting gear and calculates the stretch (the length if belt that has been pulled out) of the seat belt. This way, it can be detected whether the seat belts are applied and hence control the ignition process. Hence, it can be assured that anyone who drives any form of 4-wheeled vehicle is using a seat belt.
- **Attention Retention System (ARS):** Especially during long night journeys, the driver gets drowsy and loses attention on the road. These are one of the prime reasons of road accidents at night. ARS provides solution for these issues. A small camera module is mounted in front of the driver and is coupled with a processing board (e.g., Raspberry Pi). The camera continuously takes footage and applied suitable image processing algorithms, the processing boards detect the time duration (*t*) between the eyes are closed during a blink. If the time (*t*) increases, it means that driver is getting drowsy. After the detection, the tiny vibration motors mounted on the seat and the seat belt it vibrates to alert the driver. If the detection becomes frequent, a buzzer is set on, indicating that the driver needs to rest for a while before continuing to drive.
- **Smart Seat Belt (SSB):** This system is the combination of two main subsystems.

$$SSB = IIS + ARS$$

IIS and ARS combined give rise to the smart seat belt. It has a great potential to save thousands of lives. The trailer (truck) and bus drivers, who have the responsibility of the goods and the people they carry respectively, will get immensely benefited by this innovation. This innovation will not only save lives but also make roads safer.

- **Feasibility:** This innovation doesn't require any sort of re-design of today's vehicle. It is an add on device which can be added to any car, bus, trailer, etc., is one of the advantages.

9.3.1.6 Drunk and Drive Detection using IoT

This system is designed to detect the alcohol level in the body of the person who is driving car and avoid accidents occurring due to drunken driving. The proposed system (Figure 9.11) detects whether the person who is driving car is drunk or not. This can be done using Raspberry Pi. If the person is drunk means car is automatically slowed down or speed is reduced based on the traffic. Ultrasonic sensors are used to detect whether other vehicles are present in the front or back of the car. Based on sensor reading car speed can be varied. If any vehicle is present car is slowed or if any vehicles are not present means car is stopped. Along with drunk and drive detection drowsy driving can also be detected by using this system. This can be done using eye-blink sensor. It is used to capture the eyelids movement and if the eyes are closed for longer time vehicle's speed is controlled based on ultrasonic sensor. This will reduce the accidents. In case any accident occurs then vibration sensor confirms the accident. An alert message is sent to the family members simultaneously.

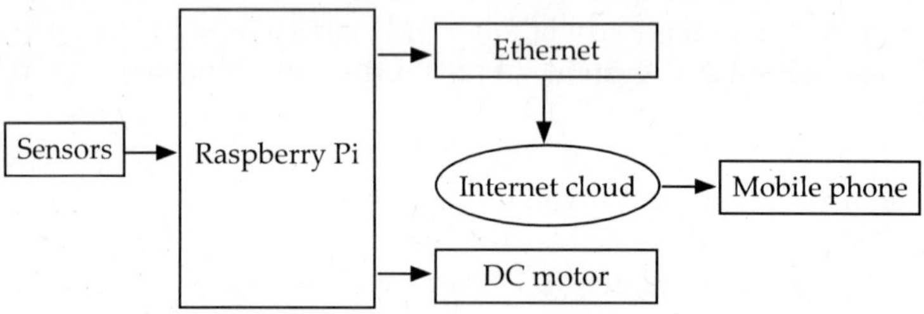

Fig. 9.11 Drunken and drive detection using IoT

All the sensors like alcohol sensor, vibration sensor, eye-blink sensor, ultrasonic sensor are connected to the Raspberry Pi. All the sensor readings are sent to the Raspberry Pi and it monitors the speed of the vehicle. DC motor is used to indicate the speed of the vehicle. Raspberry Pi is a small computer developed by the Raspberry Pi Foundation. There are two models of Raspberry Pi. They are model A and model B. The main difference between these two models is that the model B has an extra USB port and Ethernet than Model A. Various components of Raspberry Pi 7 include GPU (graphics processing unit), GPIO (general purpose input output), LEDs, USB, HDMI, power, SD card slot, Ethernet. Alcohol sensor MQ3 is used to detect the presence of alcohol in the air. It is used in breath analyzers. It has 6 pins. But only 4 of them are used. Two pins are used for heating purpose and the other two pins are used for connecting to the power and ground. It consists of SNO_2 layer inside aluminum microtubes. And there is a heating element in the tube. Ultrasonic sensors consist of two main parts called transmitter and receiver. Transmitter emits high frequency sound waves at regular intervals. They travel in air with velocity of sound. If they strike any object then their echo is received by the receiver. Distance is calculated by the time span between emitting the sound and receiving the echo. Eye-blink sensor is used to capture the eyelid movement. Eyelid movement is captured by the infrared (IR) light reflected from the surface of the eye. Vibration sensor is used to detect the occurrence of accidents. If any accident occurs then an alert message will be sent to the family members.

9.3.2 Smart Street Light

Smart street light working diagram is shown in Figure 9.12 which is used to control ON and OFF state of light based on the requirement. DHT11 sensor is used to sense the humidity. LDR sensor is used to sense the intensity variation of light. Arduino microcontroller is used to process the data. ESP 8211 (Wi-Fi module) is used to transmit the data. ULN 2003 is a relay driver capable to drive the load of 50 V and 500 mA. Relay is used to digital control switches and to glow the LED.

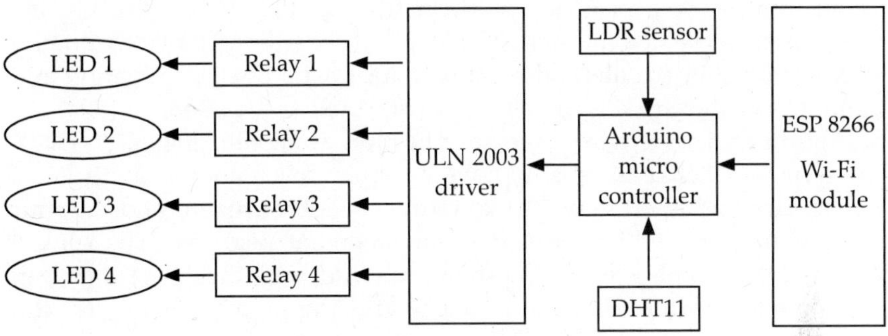

Fig. 9.12 Working of smart street light

9.3.3 Solid Waste Management

Based on the amount of waste, the waste control and management board gets the message with location on time which avoids the overfilling of waste, reduces travel charge in case of under filling and regular maintenance of garbage bin.

Arduino microcontroller (Figure 9.13) is used to process the data. ESP 8211 (Wi-Fi module) is used to transmit the data. HC-SR04 ultrasonic sensor is used to sense the depth of the bin to get the garbage level. GSM module is to get the location of garbage bin. Ultrasonic sensors measure the sound waves with higher frequencies which human beings normally do not sense. These sensors typically do not require physical contact with their target.

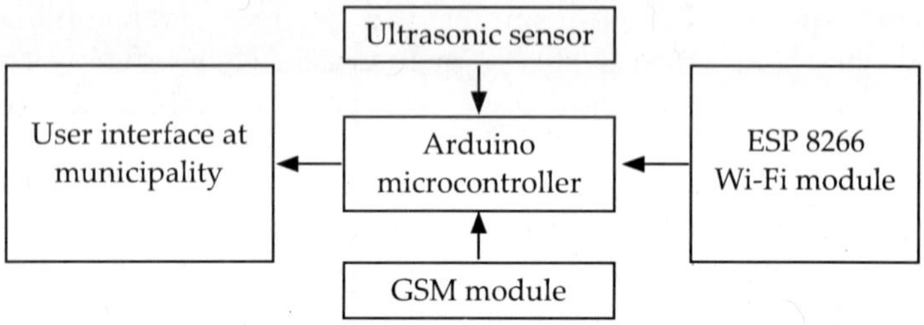

Fig. 9.13 Working of solid waste management

9.4 SMART ENVIRONMENT

The smart cities that are IoT-driven provide solutions for a healthy environment by tracking parameters and maintain them at an optimal level. For example, in monitoring water quality, a network of sensors can be deployed by the city across the water grid and connect them to a Cloud management platform. The sensors can measure pH level, the amount of dissolved oxygen and dissolved ions. When leakage occurs and the chemical composition of water changes because of water mixing with gaseous compounds, the Cloud platform triggers an output defined by the users. For example, if a nitrate (NO_3) level exceeds 1 mg/L, a water quality management solution alerts the maintenance team of contamination and automatically the workers are alerted to start fixing the issue.

Another smart environment application is monitoring air quality. Here a network of sensors is deployed in busy roads and around plants. The sensors gather data on the amount of CO, nitrogen, and sulphur oxides which are fed to the Cloud management platform. The central Cloud platform analyzes and visualizes sensor readings. The data can be accessed from the platform by users and the map of air quality can be viewed. From this data the areas where air pollution is critical can be identified and worked out. A simple block diagram that represents smart environment is given in Figure 9.14.

Public Safety

The public safety is enhanced by IoT based smart city technologies which offer real-time monitoring, analytics, and decision making tools. The data

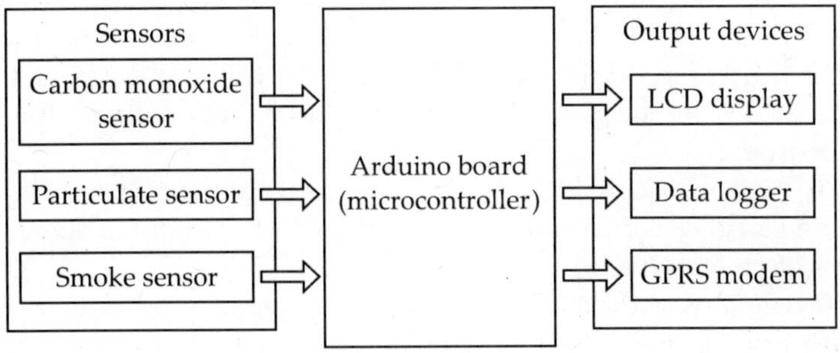

Fig. 9.14 Model of smart environment

from acoustic sensors and CCTV cameras deployed throughout the city are combined with the data from social media feed. The data is analyzed and public safety solutions can predict potential crime scenes. This helps the police to stop potential perpetrators or successfully track them.

For example, in the United States more than 90 cities use a gunshot detection system. The system uses connected microphones installed throughout the city. The data from microphones is stored in the Cloud platform. The Cloud platforms analyze, the sounds received by the microphones and detect a gunshot. The platform also measures the time it took for the sound to reach the microphone and estimates the location of the gun. When the gunshot and its location are identified, Cloud software alerts the police via a mobile app.

9.5 HEALTHCARE AND LIFESTYLE APPLICATIONS

Internet of Things (IoT) enabled devices provides a effective remote monitoring in the healthcare sector, and keeps patients safe and healthy by empowering physicians to deliver superlative care. It makes patients engagement and satisfaction as interactions with doctors in more effective way. Furthermore, remote monitoring of patient's health helps to reduce the patients need to stay in hospital. IoT also has a major impact on reducing healthcare costs significantly and improving treatment outcomes. From this, IoT is undoubtedly transforming the healthcare industry by redefining the space of devices and people's interaction in delivering healthcare solutions.

Redefining Healthcare

The proliferation of healthcare specific IoT products generated a huge amount of data by these connected devices to transform normal healthcare into smart healthcare.

IoT has a four-step architecture basically stages in a process as shown in Figure 9.15. All four stages are connected in a manner that data is captured or processed at one stage and yield the value to the next stage. Integrated values in the process bring intuitions and deliver dynamic business prospects.

Step 1: First step consists of deployment of interconnected devices that include sensors, actuators, monitors, detectors, camera systems, etc. These devices collect the data.

Step 2: Usually, data received from sensors and other devices is in analog form, which need to be aggregated and converted to the digital form for further data processing.

Step 3: Once the data is digitized and aggregated, this is preprocessed, standardized and moved to the data center or Cloud.

Step 4: Final data is managed and analyzed at the required level. Advanced analytics, applied to this data, brings actionable business insights for effective decision making.

IoT is redefining healthcare by ensuring better care, improved treatment outcomes and reduced costs for patients, and better processes and workflows, improved performance and patient experience for healthcare providers.

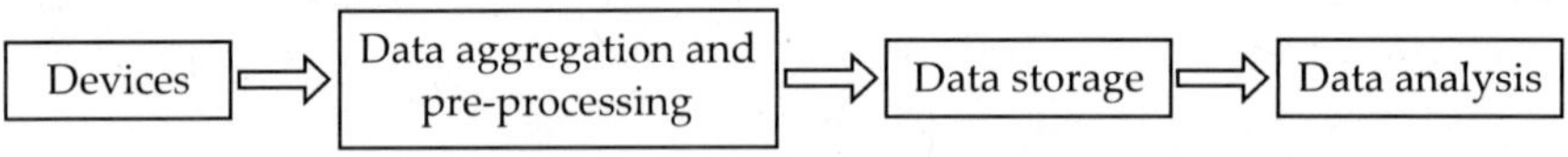

Fig. 9.15 Four stages of IoT solutions

The major advantages of IoT in healthcare include:

- **Cost reduction:** IoT enables patient monitoring in real time, thus significantly cutting down unnecessary visits to doctors, hospital stays and readmissions.
- **Improved treatment:** It enables physicians to make evidence based informed decisions and brings absolute transparency.
- **Faster disease diagnosis:** Continuous patient monitoring and real-time data helps in diagnosing diseases at an early stage or even before the disease develops based on symptoms.
- **Proactive treatment:** Continuous health monitoring opens the doors for providing proactive medical treatment.
- **Drugs and equipment management:** Management of drugs and medical equipment is a major challenge in healthcare industry. Through connected devices, these are managed and utilized efficiently with reduced costs.
- **Error reduction:** Data generated through IoT devices does not only help in effective decision making but also ensures smooth healthcare operations with reduced errors, waste and system costs.

Some of the applications of IoT in healthcare applications are as follows.

9.5.1 Smart Continuous Glucose Monitoring (CGM) and Insulin Pens

A continuous glucose monitor (CGM) system (Figure 9.16) is a device that helps diabetics to continuously monitor the blood glucose levels for several days at a time, by taking readings at regular intervals. The first CGM system was approved by the US Food and Drug Administration (FDA) in 1999, and in recent years, a number of smart CGMs have hit the market.

Fig. 9.16 Sending patients glucose value into network

Smart CGMs like Eversense and Freestyle Libre send data on blood glucose levels to an app on iPhone, Android or Apple Watch, allowing the wearer to easily check the information and detect trends.

9.5.2 Connected Inhalers

Like diabetes, asthma is a condition that impacts the lives of hundreds of millions of people across the world. Smart technology is beginning to give them increased insight into and control over their symptoms and treatment, thanks to connected inhalers as shown in Figure 9.17.

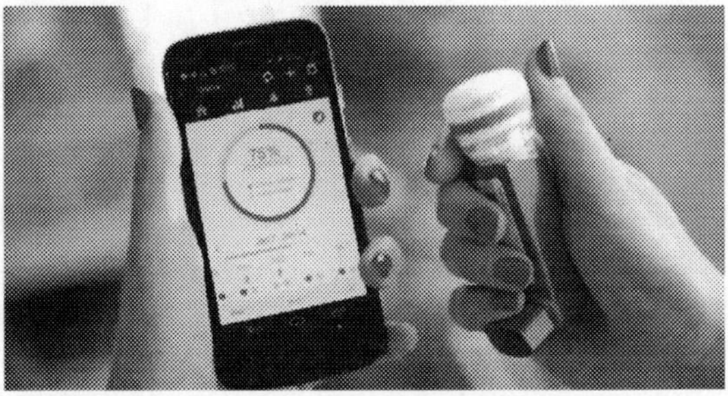

Fig. 9.17 Connected inhalers

The leading market producer of smart inhaler technology is Propeller Health. Rather than producing entire inhalers, Propeller has created a sensor that attaches to an inhaler or Bluetooth spirometer. It connects up to an app and helps people with asthma and COPD (chronic obstructive pulmonary disease, which includes emphysema and chronic bronchitis) understand what might be causing their symptoms, track uses of rescue medication, and also provide allergen forecasts.

One of the benefits of using a connected inhaler is improved adherence in other words medication is taken more consistently and more often. The Propeller sensor generates reports on inhaler use that can be shared with a patient's doctor, and it shows particular usage chart which will help to compare with prescribed. For patients, this provides motivation and also clarity, showing that how the usage of the inhaler is improving their condition.

9.5.3 Ingestible Sensors

Proteus Company has created pills that dissolve in the stomach and produce a small signal that is picked up by a sensor worn on the body. The data is then relayed to a smart phone app, confirming that the patient has taken their medication as directed. Proteus has so far trialed the system with pills for treating uncontrolled hypertension and Type 2 diabetes, and antipsychotic medication.

As with connected inhalers, ingestible sensors can help to track and improve how regularly patients take their medication, as well as allowing them to have a more informed dialogue with their physician about treatment. While the idea of taking pills with a sensor might seem invasive, the system is opt-in on the part of patients, and they can discontinue sharing some types of information, or opt out of the program altogether at any time.

9.5.4 Connected Contact Lenses

Medical smart contact lenses (Figure 9.18) are an ambitious application of the IoT in a healthcare context. While the concept has a great deal of potential, so far, the science hasn't always managed to live up to expectations.

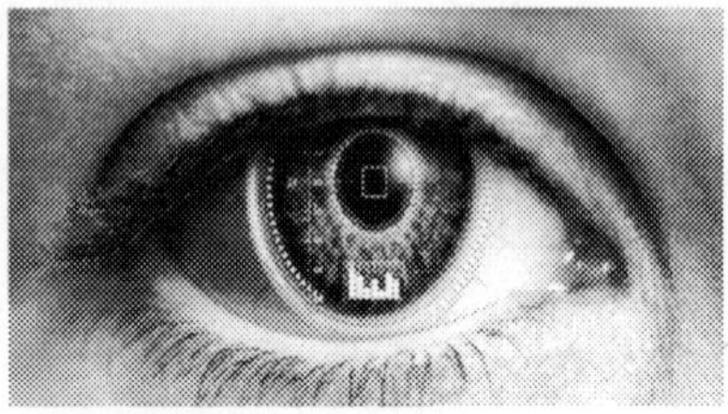

Fig. 9.18 Connected contact lenses

In 2014, Google Life Sciences (now known as Verily, a subsidiary of Google's parent company Alphabet) announced it would be developing a smart contact lens that could measure tear glucose and provide an early warning system for diabetics to alert them when their blood glucose levels had dropped or risen beyond a certain threshold. It partnered with Alcon, the eye care division of pharmaceutical company Novartis, for the project.

Smart contact lenses might prove more successful. Verily is still working on two smart lens programs with Alcon, which aim to treat presbyopia (long-sightedness caused by a loss of elasticity in the lens of an eye) and cataract surgery recovery. Swiss company SENSIMED has also developed a noninvasive smart contact lens called Triggerfish, which automatically records changes in eye dimensions that can lead to glaucoma.

9.5.5 Coagulation Testing

In 2016, Roche launched a Bluetooth enabled coagulation system that allows patients to check how quickly their blood clots. This is the first device of its kind for anti-coagulated patients, with self-testing shown to help patients stay within their therapeutic range and lower the risk of stroke or bleeding.

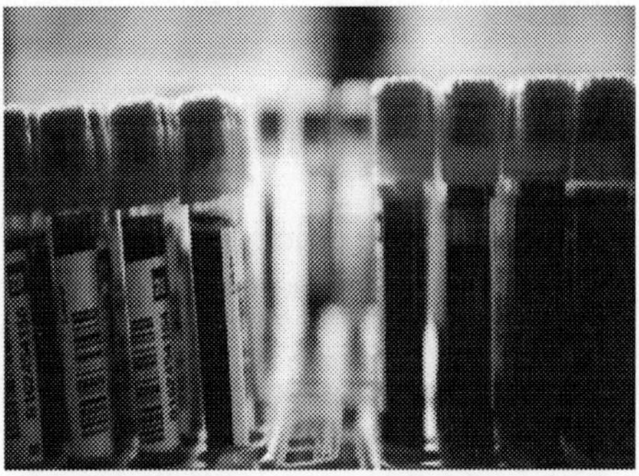

Fig. 9.19 Coagulation testing

Being able to transmit results to healthcare providers means fewer visits to the clinic. The device also allows patients to add comments to their results, reminds them to test, and flags the results in relation to the target range.

9.5.6 ADAMM Asthma Monitor

ADAMM (automated device for asthma monitoring and management) is a wearable smart asthma monitor that purports to detect the symptoms of an asthma attack before its onset, allowing the wearer to manage it before the attack gets worse.

It vibrates to notify the person wearing it of an impending asthma attack, and can also send a text message to a designated care at the same time. Other features of the device include inhaler detection the device can detect and track inhaler use, if the patient can't remember whether they've used one and voice journaling to record things like changes, feelings and behavior.

It also has an algorithm technology that learns what 'normal' is for the wearer over time, allowing it to better understand when something has changed. ADAMM works in conjunction with an app and web portal, helping asthma patients to set medication reminders, view data from the device, and remind themselves of their treatment plan.

9.6 INDUSTRIAL IoT

Industrial IoT is main component in Industry 4.0 which provide efficient outcome of productivity by this reducing the increase in demand of the product and provide good growth rate to the company. It, being the basis for the Industry 4.0 and Smart Factory, provides connectivity for smart factories, machines, industrial infrastructure, management systems, and more to streamline business operations, creating intelligent, self-optimizing industrial equipment and facilities. A number of challenges related to the management of equipment and resources, the security and safety of people can be addressed with innovative IoT solutions.

Industrial Internet of Things (IIoT) is providing with the user interface to control, sense, and analysis of data by means of it predicting the future also possible. When IIoT is needed to implement in any company the actual process which involves physical things, connectivity, control be digital twin to avoid unnecessary problem or failure

Digital twin is the virtual representation or digital replication of actual companies' physical assets, process in place, automation devices and system. The twin provides the real time data to enable learning and reasoning for improved decision making. It allows experimentation with new information that is generated by Cloud based artificial intelligence function.

For example, Caterpillar is well known for mining equipment manufacturing they have around 10,000 fleets around the world. They provide the 2000 fleets without any connectivity to device after integration all the fleets with IIoT their growth rate increase by 10-15 percentage. By providing this they know the status of machine, working area, and its efficiency by this they provide great customer support.

Table 9.1. IIoT vs IoT

Based on factors	IIoT	IoT
Application area	Its concentrates on connection of machines and devices in industries like manufacturing, healthcare and logistics.	It is commonly used to concentrate on customer based devices like fit bits and wide range of smart home devices.
Safety	In these connected equipment cause life threatening or emergency situation in the case of failure.	It is used on the common devices so that it causes any life threatening or emergency situation.
Security	Cyber security is mandatory.	Cyber security is mandatory.

Process	Cloud computing with AI is needed	Cloud computing along is sufficient
Timeliness	Mandatory	Not mandatory
Implement cost	High	Comparatively low
Life cycle	High	Short
Maintenance	Needed	Not needed
Requirements	Stringent	Moderate
Network	Large scale	Small scale

9.6.1 Applications

(i) ABB – Self-balancing Robot

It's mainly preferred in industry to lift the material from one place to another inside industry which enables the connectivity to track them and utilize it efficiently. Figure 9.20 gives a model of self-balancing robot for an easy understanding.

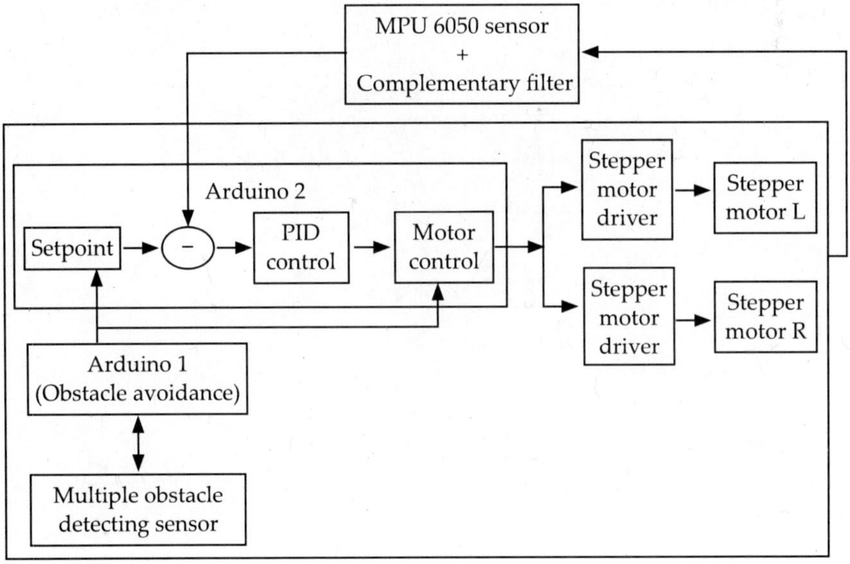

Fig. 9.20 Self-balancing robot

(ii) Hitachi

It is a Japan based company. Figure 9.21 represents how they implemented digital supply chain by IIoT concept.

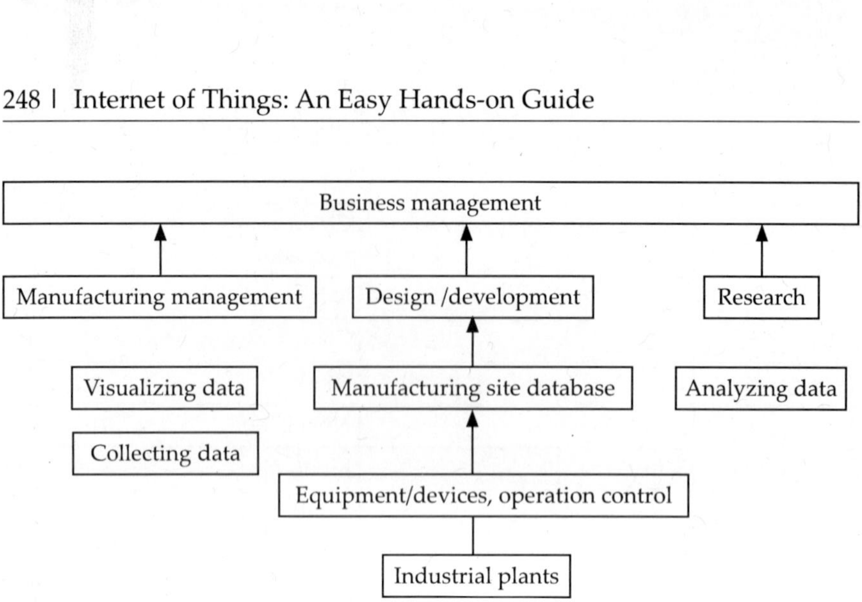

Fig. 9.21 Digital supply chain using IoT

Industry 4.0 is at its peak. Every industry in the market is evolving into a smart industry. With the implementation of smart solutions in business processes, organizations are leveraging the benefits of IoT and elevating their businesses to the next extent. The foundation of a smart industry can be laid upon three pillars given in Figure 9.22.

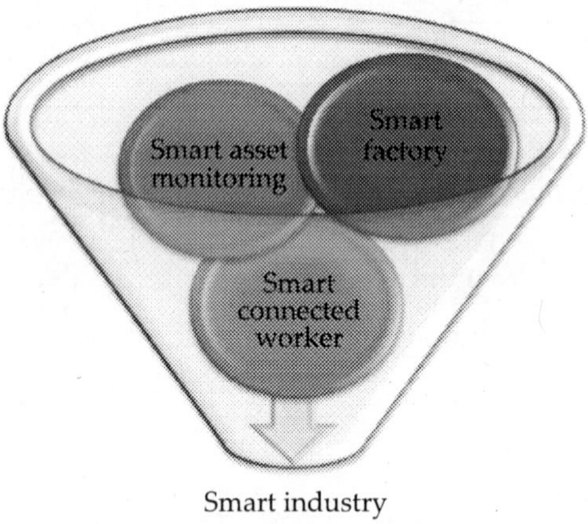

Fig. 9.22 Blocks in smart industry

9.6.2 Smart Factory

The key challenges faced by industries are:

- Lack of real-time data and disparate data systems
- Keeping abreast of regulations and managing compliance
- Implementing procedures that allow faster time to market
- Balancing maintenance with throughput
- Ensure customer satisfactions

With a smart factory solution, supervisors get real-time, continuous and standardized task level data to perform a detailed analysis of factory performance. With a better knowledge of their production lines, supervisors can easily make out the problematic areas and take informed decisions to meet the dynamic demands of production lines.

9.6.3 Smart Asset Monitoring

Prior to smart asset monitoring solution, organizations were facing challenges like:

- Tracking asset performance
- Limited control over assets
- Loss of assets
- Separate silos of IT and OT
- Scheduling an effective maintenance plan

An IoT-enabled smart asset monitoring solution not only helps in gaining better control over the equipment and machinery of a business but also provides proficiency by increasing efficiency and productivity, reducing operational costs, increasing customer experience, etc. With the help of a smart asset monitoring solution, businesses can get deeper insights about health, availability, productivity, and location about the assets. What makes the solution so popular is the ability to monitor all the above-mentioned attributes from a single point of view. Also, the feature of keeping the record of every active asset, eliminating any ghost assets and providing intact information on how assets are being utilized in every process adds the cherry on the cake for a smart asset monitoring solution.

Smart Connected Worker

Key challenges faced by industries are:

- Unexpected accidents in the field
- Improving worker performance
- Enhancing employee experience
- Track the locations of workers and reducing the risk of errors
- Improving data accuracy
- Creating visibility in the process work status

Connecting industrial workers, in addition, to have a fully connected system has now become a feasible solution to achieve better operational efficiency and to ensure worker safety and productivity. Along with connected assets and machinery, connected workers can help a company gain better regulatory compliance and take a real-time glance at a worker's activity and health.

Why do industries need these smart solutions?

Stakeholders may face challenges like:

- Production line breakdown
- Unavailability of assets
- Loss of final product
- Injured labor
- Too many resources on one task

Let's have a look at the benefits an organization can avail by implementing these three solutions together.

(a) **Improved operational efficiencies:** Smart industrial IoT solutions use real-time data from sensors to monitor and enhance process efficiency, reduce human intervention, and energy costs. All data obtained can be utilized to minimize downtime due to maintenance, unavailability of assets and lack of human resources in the factory.

(b) **Better safety and security:** Besides resolving situations faster by coordinating with multiple teams, these solutions allow businesses to improve workplace safety and guard against physical threats using sensors and video cameras that help monitor equipment.

(c) **Faster decision making;** Organizations can use real-time operational insights to make faster and smarter business decisions and reduce operating costs. These IoT enabled solutions help distribute intelligence and prioritize all business decisions.

Smart solutions can serve as a forecasting tool for better decision-making by constantly monitoring, aggregating, and analyzing data from various sensors installed across the factory or deployed to the entire staff. These sensors help supervisors to collect real-time data, which in turn, allows companies to obtain insights into various operations and initiate preventative actions.

To be in the league of growing businesses, getting smart can be an optimal solution for your business. By incorporating smart industrial solutions in a business, enterprises can convert real-time data into actionable insights, which in turn, can help in better decision making.

9.7 CASE STUDIES

9.7.1. Plant Health Monitoring System Using Raspberry Pi

Plant health monitoring is one of the most important tasks in any agriculture based environment. Thus, an efficient monitoring system is required for continuous and long-term monitoring. This work aims at monitoring the plant's health based on NDVI (normalized differential vegetation index) which helps in recognizing the difference between healthy and an unhealthy plants by calculating their NDVI values. The images of the plants are taken through the NIR camera which is interfaced with the Raspberry Pi. The Raspberry Pi is coded with Python to capture images and image processing technique is used to calculate the NDVI values. This also includes the acquisition of parameters like soil temperature, soil moisture, humidity and soil pH level using appropriate sensors. The final step is to develop an Android application using Android Studio software. The processed data is transmitted to the terminal user. It allows the terminal user to access the data, achieve a timely interaction and cure of the intended crop. The proposed system operation is validated with an experimental setup.

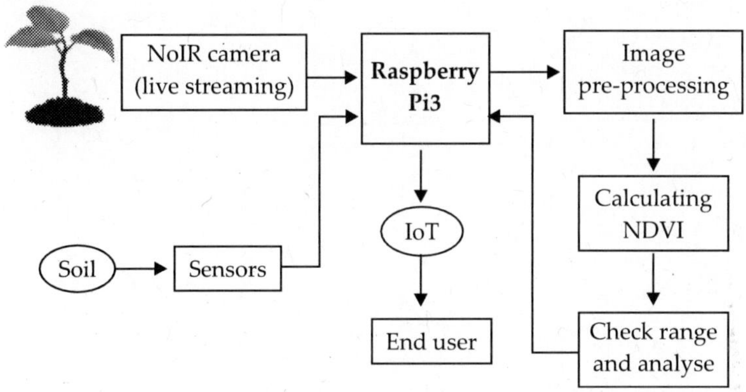

Fig. 9.23 Block diagram of plant health monitoring system using Raspberry Pi

Figure 9.23 shows a vivid picture of the implementation of the system. Here the data from each block are directed to the microcontroller for processing the data into useful information. This information is directed to the Cloud server and all the information sorted in order for storage. This information can be collected by user any time in his/her smart phone.

Reference code for sending data from sensors to thinkspeak

```
import sys
import urllib.request
from time import sleep
import Adafruit_DHT as dht
# Enter Your API key here
myAPI = '2HVZO4UBA4PUJ5PU'
# URL where we will send the data, don't change it
baseURL = 'https://api.thingspeak.com/update?api_key=%s' % myAPI
def DHT22_data():
```

```
# Reading from DHT22 and storing the temperature and humidity
humi, temp = dht.read_retry(dht.DHT22, 23)
return humi, temp
while True:
try:
humi, temp = DHT22_data()
# If Reading is valid
if isinstance(humi, float) and isinstance(temp, float):
# Formatting to two decimal places
humi = '%.2f' % humi
temp = '%.2f' % temp
# Sending the data to thingspeak
conn = urllib.request.urlopen(baseURL + '&field1=%s&field2=%s' %
(temp, humi))
print (conn.read())
# Closing the connection
conn.close()
else:
print ('Error')
# DHT22 requires 2 seconds to give a reading, so make sure to add
delay of above 2 seconds.
sleep(2)
except:
break
```

For other transmission refer Python code in

https://github.com/PiddePannkauga/ndviMachine

9.7.2 Plant Health Analysis Using Various Techniques

The new era in communication network is Internet of Things (IoT), gaining importance because of a wide variety of application in oriented projects developments. The IoT is furnishing people with smart and remote approach, the remote applications such as: smart agriculture, smart environment, smart security and smart cities, etc. These are the upcoming technologies making the things easy. The main components of IoT are sensors, actuators, embedded systems and Internet connection. Nowadays, smart agriculture needs to improve the cultivation methods and increase harvesting with good quality. Based on these aspects, sensors are available for specific agriculture application, like soil preparation, crop health status, irrigation, insect and pest detection. Plant diseases diagnosis using sensors and image processing is planned in this project. The irrigation level and health monitoring is measured by different sensor. Plant diseases diagnosis and leaf region affect diseases identification using image processing techniques. Plant leaf diseases using identification image enhancement and image restoration using leaf region noise removal and quality in the plant.

9.7.3 IoT based Wearable Tech with Health Monitoring

Smart watch is the most significant item in wearable device category. The phone that has been picked out of the pocket and now is on hands, a person still needs to do some steps such as unlocking the phone, finding the app. The number of steps involved is high and thus a smart watch device is helpful. Smart watches that are available in market are of high cost most of us can't afford it. The home automation modules like fans, lights can be controlled using this watch by connecting to Wi-Fi. The SMS and call notifications are pushed into the screen of the watch. It also keeps track of heart pulse rate and get connected to Bluetooth. This watch is programmed using Arduino's open source libraries and functions. This watch is embedded in heart rate sensors, Node MCU and Bluetooth module that track and interpret user's action and information. This has two mechanisms. One that consists of sensors embedded in the circuitry of watch whereas the second mechanism is the interface between a smart phone and the watch. This can be altered as per user needs. Figure 9.24 represents the full process.

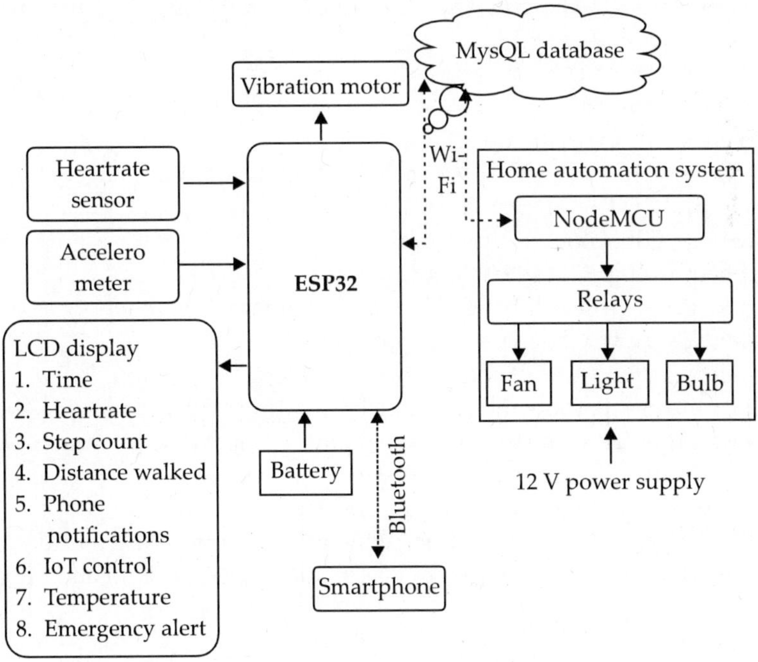

Fig. 9.24 Block diagram of IoT based wearable tech with health monitoring

Code for IoT based Wearable Tech with Health Monitoring

Sofware Arduino IDE

```
//Tranceiver Code
    String LoRaData = LoRa.readString();
    Serial.print(LoRaData);
    }
    // print RSSI of packet
    Serial.print(" with RSSI ");
    Serial.println(LoRa.packetRssi());
```

Interfacing ESP8266 NodeMCU with Nokia 5110 LCD (84×48 pixel).

This is a free software with NO WARRANTY.

https://simple-circuit.com/

```
*************************************
```

This is an example sketch for our Monochrome Nokia 5110 LCD Displays

Pick one up today in the adafruit shop!

------> http://www.adafruit.com/products/338

These displays use SPI to communicate, 4 or 5 pins are required to interface

Adafruit invests time and resources providing this open source code, please support Adafruit and open-source hardware by purchasing products from Adafruit!

```
***************************************
#include "time.h"
#include <SPI.h> // include SPI library
#include <Adafruit_GFX.h> // include adafruit graphics library;
```

Adafruit MQTT Library ESP8266 Example

Must use ESP8266 Arduino from:

https://github.com/esp8266/Arduino

Works great with Adafruit's Huzzah ESP board & Feather

----> https://www.adafruit.com/product/2471

----> https://www.adafruit.com/products/2821

Adafruit invests time and resources providing this open source code, please support Adafruit and open-source hardware by purchasing products from Adafruit!

ritten by Tony DiCola for Adafruit Industries.

MIT license, all text above must be included in any redistribution

```
// Setup the MQTT client class by passing in the Wi-Fi client and MQTT server and login details.
Adafruit_MQTT_Client mqtt(&client,
AIO_SERVER, AIO_SERVERPORT,
```

```
AIO_USERNAME, AIO_KEY);
// ping the server to keep the mqtt connection alive
// NOT required if you are publishing once every KEEPALIVE seconds
/*
// Function to connect and reconnect as necessary to the MQTT server.
// Should be called in the loop function and it will take care if
connecting.
mqtt.disconnect();
delay(5000); // wait 5 seconds
//pretty simple function - read everything out of the serial buffer and
whats coming and get rid of it
void serial_dump_ESP(){
char temp;
while(ESP8266.available()){
temp =ESP8266.read();
delay(1);//could play around with this value if buffer overflows are
occuring
}//while
//Serial.println("DUMPED");
}//serial dump
the ESP
if(read_until_ESP(keyword_OK,sizeof(keyword_OK),5000,0))//go
look for keyword "OK" or "no change
Serial.println("ESP CIPMUX SET");
else
Serial.println("ESP CIPMUX SET FAILED");
serial_dump_ESP();
//that's it! Could be done by nesting everything together, so if one
thing fails, it returns '0', and if it gets all the way through it returns
'1'...oh well
}//setup ESP
```

9.8 SUMMARY

The various domains of applications were selected and built IoT model with python code, controller board, API and cloud interfaces.

REVIEW QUESTIONS

1. Summarize the components needed to design home automation.
2. Construct a model of to develop a smart city.
3. What is IIoT? Explain it with some examples.

4. Describe any two applications of IoT in healthcare.
5. Does IoT development supporting agriculture?
6. How a normal factory can be changed into a smart factory? Draw the flowchart of its steps.
7. Write the working principle of smart asset monitoring.
8. How does smart watch support to human health monitoring?
9. Is it possible to run the IoT applications without Internet?
10. Write the advantage of smart meter.
11. Is a smart lock really a secured one?
12. What is the output voltage of a sensor?
13. How do you define the quality of the device life?
14. Vehicle speed checker is a digital or analog device.
15. List the components which make air pollution.
16. What is meant by solid waste management?
17. What is ADAMM?
18. List the blocks of smart industry.

Annexure 1
Setting up of Arduino Experiments

EXAMPLE 1: Smart Irrigation System

The main objective is to provide revolution in agriculture by implementing an automatic plant watering system using Node MCU. It senses the moisture level of the soil and it gives a notification to our smart phones. Based on the received information, smart system will decide whether the plant must be watered or not. According to the notification, relays are used to turn ON/OFF the water pump from a remote location with help of a mobile app (Blynk app).

Module Setup

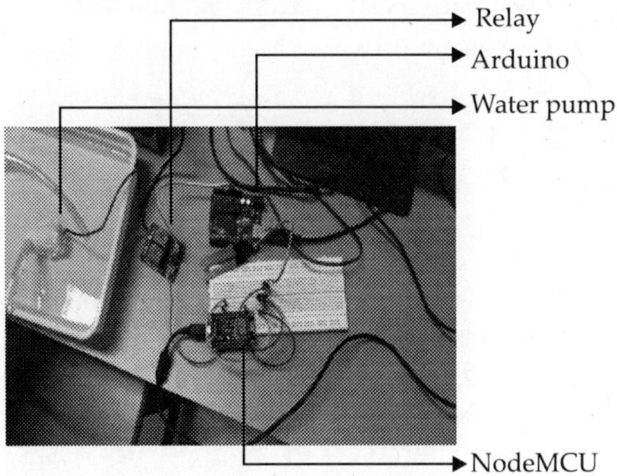

Fig. A1.1 Module 1

Procedure

Figure A1.1 Module 1 represents the setup of smart irrigation system. In this, circuit connections are given based on the default pin procedure of the specific boards. Working of application mainly depends on the sensor data. The range to find the soil dryness and wet value was referred to and stored as a threshold value in the Node MCU. Once the particular value is reached, the water pump is either automatically turned ON or manually can control it using the mobile phone blynk app after receiving notification. Blynk app can simultaneously receive notifications and provide manual control using switches in it. The output notification obtained in the blynk app is given in Figure A1.2. After receiving this notification people can turn motor on and off by pressing buttons in blynk app for fully automation system or can do by manual.

Outputs

Fig. A1.2 Bylnk app output screen

EXAMPLE 2: Emergency Alert System

The effective device-to-device (voice and data) transmission is essential for all times. The transmission was affected by many factors including transmit power, antenna characteristics, receiver sensitivity, system cost along with range limitations. Range limitations define the usable coverage area for specified radio communications. If the range of a radio system coverage area decreases, then it causes network issues and sometimes completely disconnected from the existing network. During any natural disaster, people mostly suffered with no network issues especially couldn't inform any basic need or couldn't do emergency messages. For that kind of problem would be solved by proposing LoRa based emergency alert system. This system is to increase range limitations in wireless communications like the transmit power, the gain of transmit and receive antenna and to increase receiver sensitivity. LoRa basically creates a local base station with maximum of 5 km coverage. It also provides Wi-Fi access to that person under this umbrella. So they can send either voice or text by this temporary network which having low bandwidth during the emergency period.

Module Setup

Fig. A1.3 Module 2

Procedure

The total working procedure connected in the module 2 can be classified into 4 main phases. They are

1. Setup phase,
2. Getting input message to be transmitted,
3. Transmission, and
4. Received message as output.

Once any emergency situation arose just pressing the keypad keys can send our important needs (already assigned templates for each key). The receivers within the range of around 5 km from the transmitter distance can use this facility. The output obtained from the LoRa transceiver is given as Figure A1.4.

Outputs

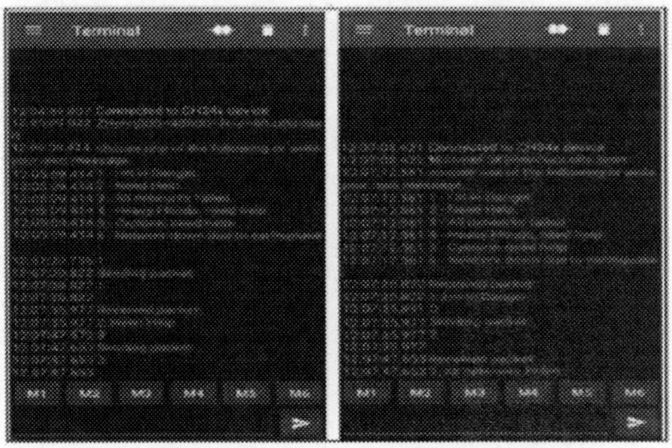

Fig. A1.4 Transmission and reception screen

EXAMPLE 3: Fallen Power Line Detection

Features of IoT can be employed for power line fallen detection and also can create an alert system to both the electricity board and admin of snapped power lines is designed using NodeMCU. The system uses LoRa (long range), a low power wide area network (LPWAN) technology which improves the longevity of the embedded system. LoRa is battery operated and covers a transmission range of about 5 km in rural areas (line of sight). This is achieved by a trade-off with the data rate. The data rate of LoRa is low and is suitable for sensor data transmission. Hence, LoRa finds application in emergency communication system (ECS) which is connected with 'Thingspeak' network, an open IoT platform which can be easily accessed and user friendly.

Module Setup

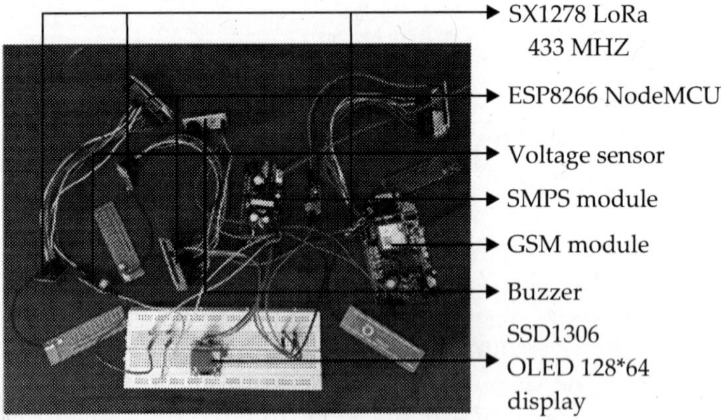

Fig. A1.5 Module 3

Procedure

A wireless sensing network based on IoT technology is demonstrated in pursuit of a monitoring, detecting and alert solution for fallen power lines. At first the technical requirements on the essential units of the IoT sensing network including integrated sensor node, power management and LPWA communication is designed. Then a general framework of the wireless sensing network including sensing, network, data and application layers is designed in ThingSpeak for communication extendibility and accessibility. As per module setup, connections are made. The voltage drop across each post was monitored and recorded. If any power line drops identified then immediately sent alert message to EB office and also the power line becomes open circuit. The designed system detected as well as alerted the users with the fallen line details are shown in Figure A1.6.

Outputs

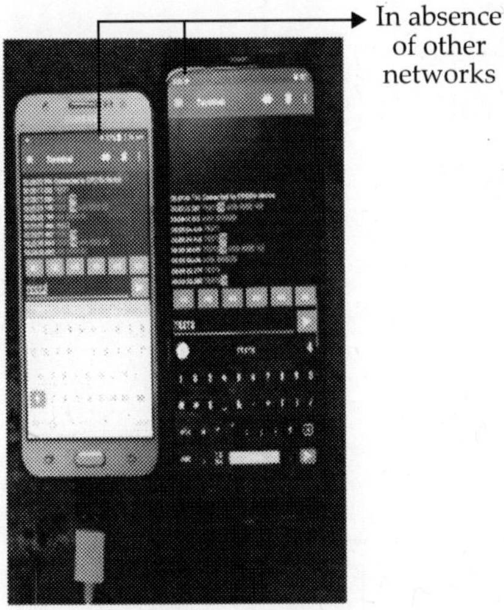

Fig. A1.6(a) Output screen

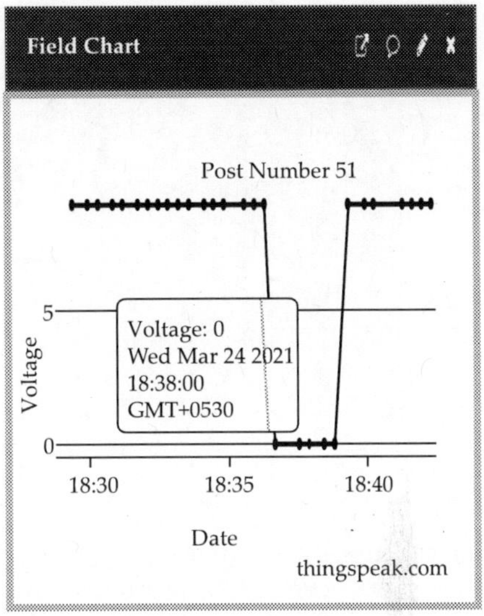

Fig. A1.6(b) Voltage drop

EXAMPLE 4: Smart Health Monitoring Watch

Smart watch is the most significant item in wearable device category. If people use smart phones for doing particular operations, the phone that has been picked out from pocket and now is on hands, a person still needs to do some steps such as unlocking the phone, finding the app for doing specific operations. The number of steps involved for this process is high and also it is tedious in case of emergency situations. Thus, a smart watch device is helpful for automatic detection alert. Smart watches that are available in market are of high cost mostly common people can't afford it. But, the newly designed watch is cost effective and has some in-built device control operations. The home automation modules like fans, lights control is in-built in that watch and it can be controlled using an existing home Wi-Fi connection. The SMS and call notifications are pushed into the screen of the watch itself. There is no need of any additional applications for this. This watch also keeps track of heart pulse rate and gets connected to Bluetooth. It is programmed using Arduino's open source libraries and functions. This watch is embedded with heart rate sensors, node MCU and Bluetooth module that track and interpret user's action and information. This proposed smart watch has two mechanisms. One that consists of sensors embedded in the circuitry of watch whereas the second mechanism is the interface between a smart phone and watch. The application mechanism can be altered as per user needs.

Module Setup

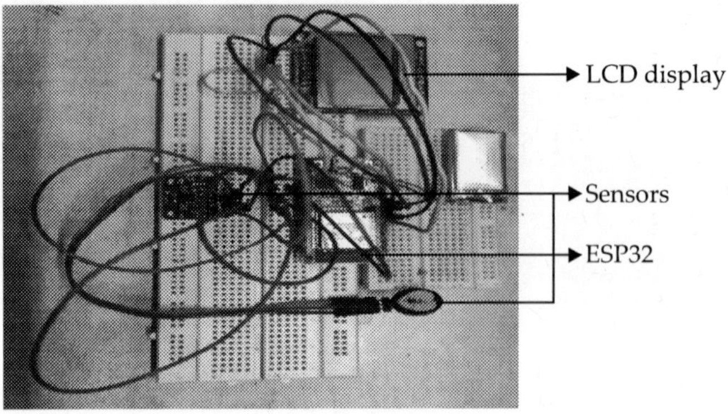

Fig. A1.7 Module 4

Procedure

When the module setup is powered up, the ESP32 gets booted and connects to the Wi-Fi network. The Bluetooth module of ESP32 gets paired with the smart phone of user. The LCD screen displays a menu that contains different options like heart rate (BPM), total steps, temperature, time and home automation controller. When the user selects the heart rate option, the pulse sensor gets powered up and starts to sense the heartbeat and displays the BPM value. When the user selects the steps count option, the accelerometer gets powered up and starts to sense the count value.

When the home automation control is selected, the ESP32 connects with the MySQL database through Wi-Fi. Then the user selects the device to be turned on or off. According to the user's input, the microcontroller adds data to the string and passes it on to the website. A webpage is created to view and update the database, so that the appliances can be turned on and off not only using the wearable tech, but also using the website. The outputs obtained for each automation function is given in Figure A1.8.

Output

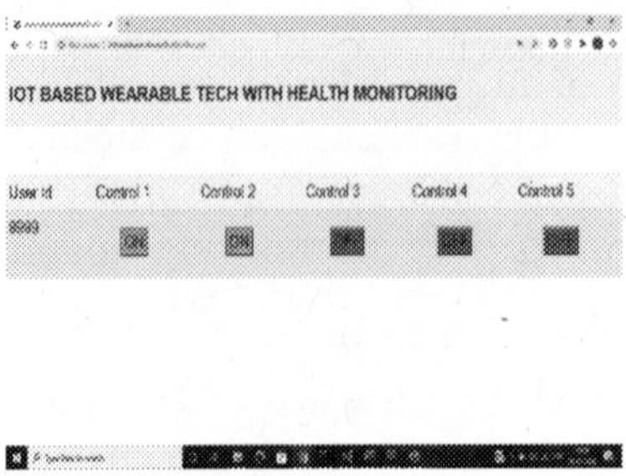

Fig. A1.8(a) Website creation

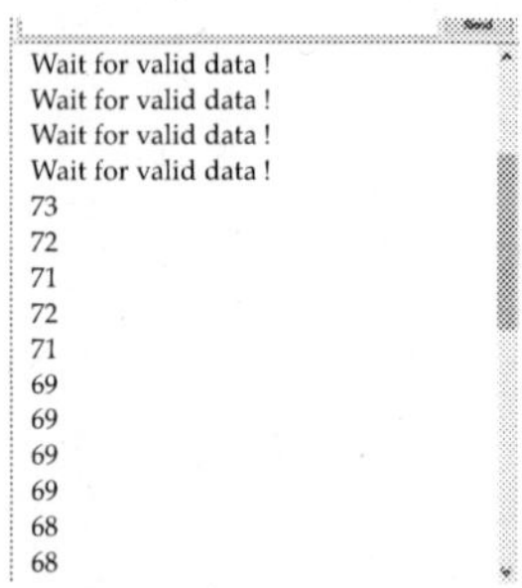

Fig. A1.8(b) Pulse sensor readings

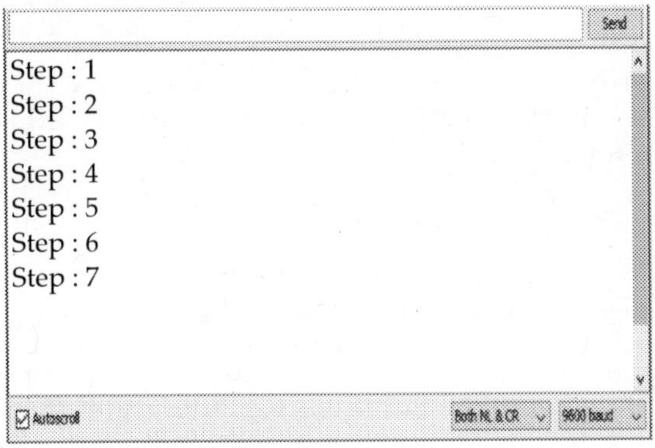

Fig. A1.8(c) Steps count readings

EXAMPLE 5: Braille Decoding System

Despite all the advances in the field of telecommunication visually impaired people have no access to modern technologies. This system proposed to bridge the gap between the blind users and the end users in the telecommunication field Braille decoding system was planned to make visually impaired can communicate and interact with others by providing smart keypad.

Module Setup

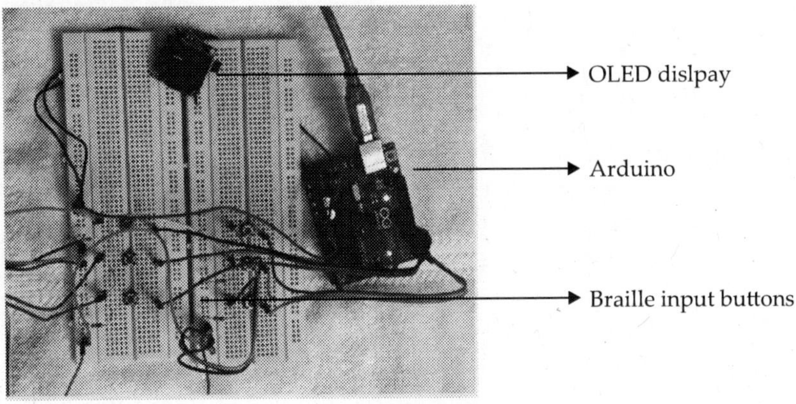

Fig. A1.9 Module 5

Procedure

This module represents the design and implementation of enhanced Braille keypad. Braille input is analyzed and created a lookup table by having multiple push buttons as input and mapped into the Alphabet character. After converting to text the output is displayed in OLED. The Braille alphabet is represented as in Figure A1.10. Based on that the Braille input was given and the corresponding text is displayed in the OLED screen and also possible to send the mobile.

Output

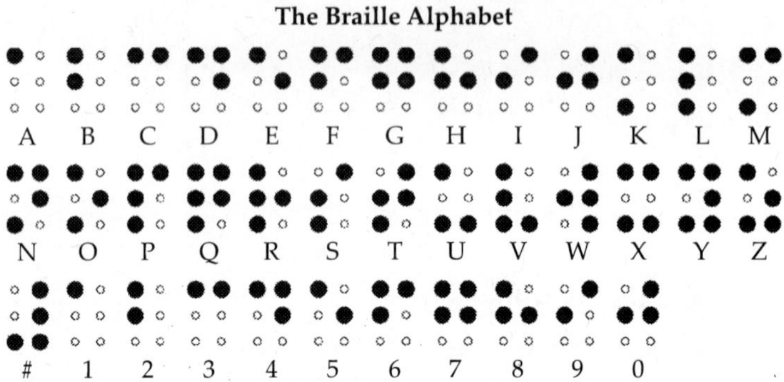

Fig. A1.10 Braille code for alphabets.

(https://www.pharmabraille.com/wp-content/uploads/2015/11/braille-alphabet-and-braille-numbers.png)

Annexure 2
Setting up of Raspberry Experiments

EXAMPLE 1: Animal Detection and Smart Alert System

Design of a detection and alert system can act as a virtual barrier covering elephant corridors or villages. Human-elephant conflict has been increasing and remains unsolved. A prototype system to detect elephants and send a warning message to the authorities and the villagers has been implemented. At present electric fencing is the main solution for human elephant conflict which is not reckoned as a sustainable solution. Need to have an optimum design which can minimize the above-mentioned limitations and to create harmony between two communities. This is a virtual barrier located outside the village and therefore can be used as an early warning system. Mainly as a vigilant method, mainly focuses to minimize the human-elephant interaction and finally chase away the elephant by using standard methods.

Module Setup

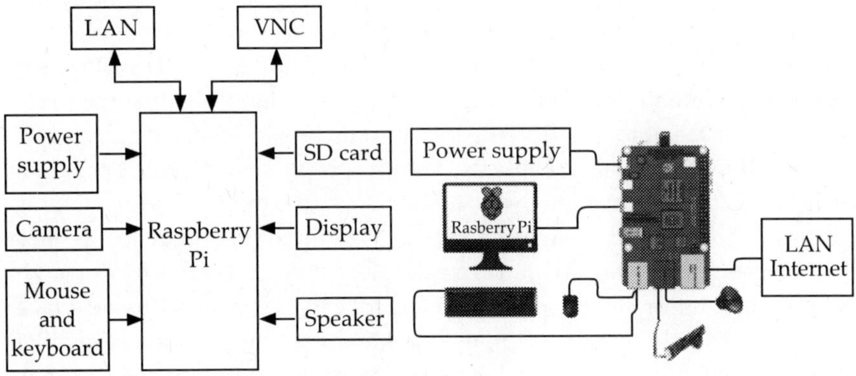

Fig. A2.1 Module 1

Procedure

Module 1 in Figure A2.1 system developed with YOLOv3 (You Only Look Once). It is a pre-trained Deep learning model for object detection. Captured image divides the image into a 13 × 13 grid of cells and image analyzed by algorithm. Once the elephant entry is identified by algorithm, alert message is passed to respective person and providing flash light with sound to run through the elephant.

Outputs

Fig. A2.2 Elephant detection using deep learning technique

EXAMPLE 2: Analysis of Fruit Ripeness Based on IoT

With the exponential growth of world population, the food production and farming needs to enhance the farm yield. Nowadays, technology is constantly improving to solve human problems more easily. The scope for manual experimentation, viability assessment through trial and error, etc., are no longer feasible. The intelligent system can be developed for fruits that can allow the quality analysis. The proposed system will provide a much better and convenient way for farmers to monitor the fruit. This includes a series of steps from capturing the image of fruit to identify whether the fruit is fully ripe or moderately ripe or immature through the implementation in Raspberry Pi. Raspberry Pi is used to interface the camera and the captured images are analyzed by various steps like acquisition, pre-processing, and segmentation. The status of the fruit has been sent by GSM module through SMS and displaying the status on the monitor display of the owner of the system. This smart way of analyzing the fruit will enable growers and farmers and helps them to understand the potential of IoT market for agriculture by installing smart technologies to increase competition and sustainability in their production.

Module Setup

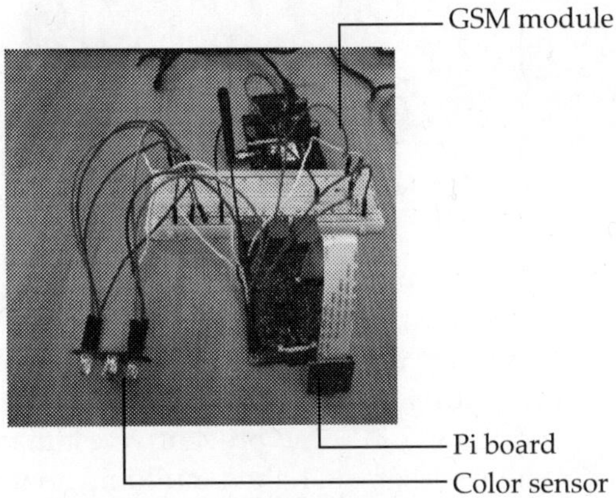

Fig. A2.3 Module 2

Procedure

Figure A2.3 module 2 shows the top view of the setup. In that fully, moderately ripened and raw lemon fruit placed in front of camera module. The rectangular region of captured fruit image using Grabcut image segmentation algorithm was obtained. Then the morphological operation of the segmented image using HSV color mask for green to yellow values and based on these color masking, contouring was done to detect the ripeness state. The RGB value of the input image was measured using color sensor. The final status of the fruit is identified based on the fruit color. The obtained result is transmitted to the user mobile phone via GSM module are given in Figure A2.4.

Outputs

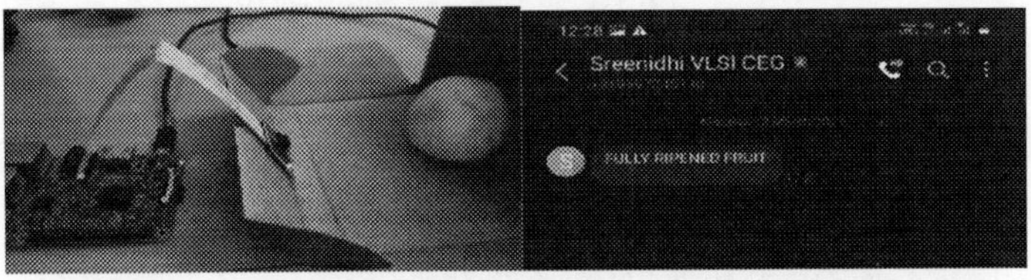

Fig. A2.4 (a) Detection of fully ripened fruit

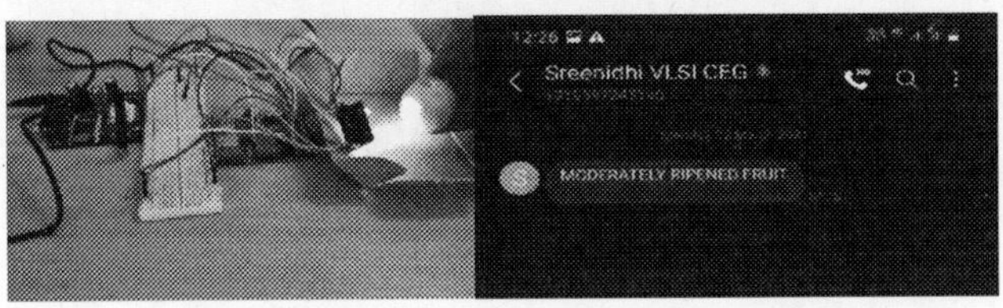

Fig. A2.4 (b) Detection of moderately ripened fruit

Fig. A2.4 (c) Detection of raw fruit

EXAMPLE 3: Smart Antipoaching Alert System

Nowadays numerous occurrences about pirating of trees like Sandal, teak and so forth are common. These trees are expensive, utilized in the medicinal sciences and beautifying agents. To limit their sneaking and to spare woodlands around the world some preventive estimates should be conveyed. IoT based remote sensor can help to build up a vitality effective framework for checking the poaching of trees. Along with that required sensors are connected with routing protocol in wireless senor networks are responsible for forwarding the information within the network. Smuggling of trees in timberland causes monetary harm and also creatures are losing their regular habitat. Raspberry Pi based used system to identify enemy of poaching framework give deal with remote sensor arrange innovation, which is practiced of detecting burglary by checking the sign delivered by cutting of trees utilizing tilt sensor, temperature sensor, metal sensor, webcam and sound sensor. The wireless sensor network is widely cast off technology in monitoring and controlling for the remote application.

Module Setup

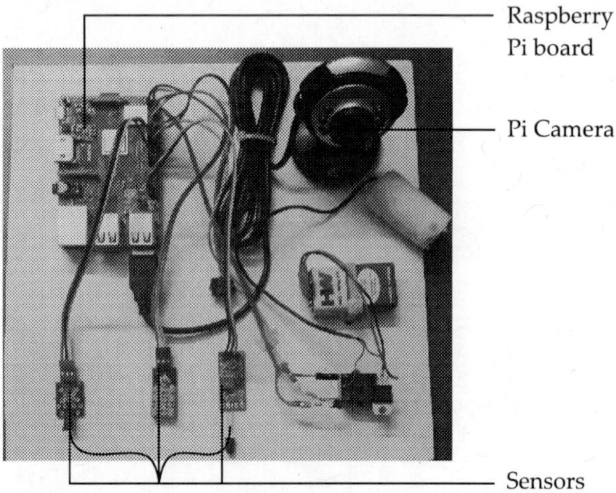

Fig. A2.5 Module 3

Procedure

Module 3 in Figure A2.5 designed to easily track the poaching activity. It uses various sensors to detect the vibration and the angle of tree while it is being cut. When an abnormal condition is noticed, GSM immediately sends a message to the base station so that the authorized person can take necessary action. Image processing is used for live video streaming, image capturing using camera.

Outputs

Apart from the main objective, the system also helps in reducing the global warming to a great extent. The natural habitat of the animals is preserved indirectly. The trees can also be protected from forest fire by using this system. This, in turn, helps in reducing deforestation. Thereby, the ecological balance is maintained. The main purpose of this project is to protect the valuable trees such as sandalwood, teakwood, rosewood, etc.

EXAMPLE 4: Smart Attendance System

In this attendance system RFID cards are used to read the name of the person and store the data in the MySQL database. Raspberry Pi is used to carry out the whole process. This system was designed in such a way that it was suitable for schools, colleges, and companies which use the attendance system.

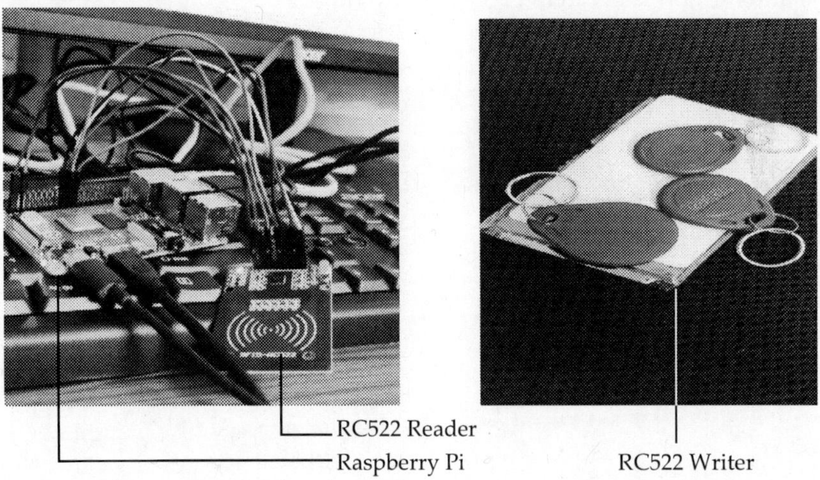

Fig. A2.6 Module 4 with RC522 Reader and Writer tag

Procedure

Figure A2.6 module 4 for attendance system connecting with RC522 RFID tags and reader were used to store the individual student information.

Then each student has to show his/her tag to the reader for the data entry. Once the tag is shown the data is collected and displayed in the open source MySQL database. In addition to each student's name, the time of arrival and time of departure of the student is also identified. The output, displayed in the MySQL database, is given below.

Output

Fig. A2.7 Attendance database in MySQL

EXAMPLE 5: Smart Automation

Nowadays, most of the people want the automation in their day-to-day life. Smart living makes the life of human easier and faster. In this work smart home automation is used to control home appliances automatically by using two different sensors. The sensors sense the environment and do the control operations as per needs. The control operation can be made both automatic and semi-automatic as per the existing electrical and electronic gadgets and the need of the users. The module setup for the home automation is given in Fig. A2.8 (module 5).

Module Setup

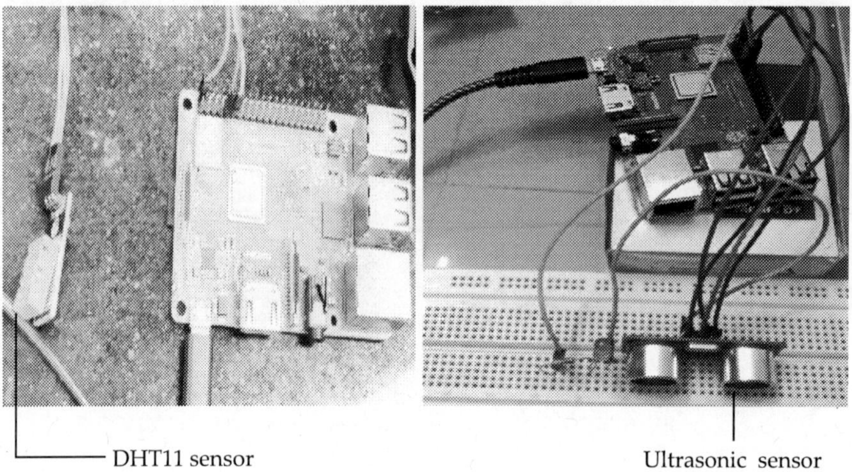

(a) (b)

Fig. A2.8 Module 5

Procedure

DHT11 sensor is used to sense the temperature and humidity. According to the acquired rate can turn ON/OFF air conditioning systems either remotely (by using remote mobile apps) or automatically and also control fan switches. Interoperability is the major challenge in the IoT devices. Smart home automation devices designed here are modelled in such a way that it can either control devices automatically or manually. In this work, ultrasonic sensor are used to identify the overflow in the water tank and control motor system either remotely or automatically. The output got from both the modules was given in Fig. A2.9.

Outputs

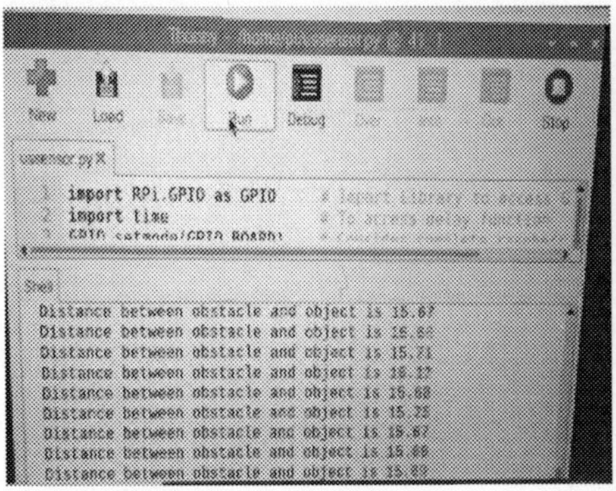

Fig. A2.9(a) Output of ultrasonic sensor

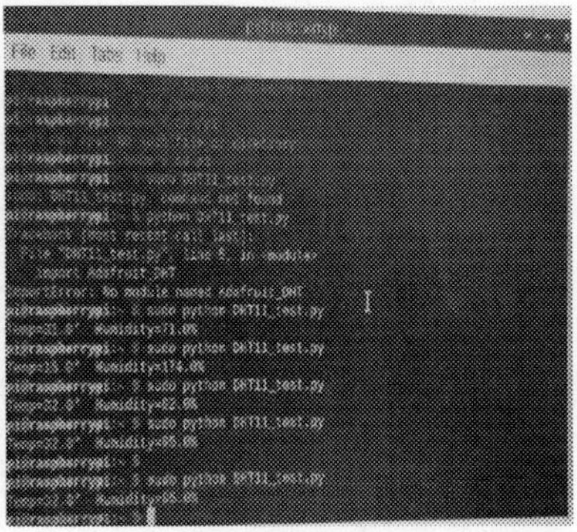

Fig. A2.9(b) DHT11 sensor

INDEX